DEEP TIME

DEEP TIME

..

How Humanity Communicates Across Millennia

GREGORY BENFORD

Perennial

An Imprint of HarperCollinsPublishers

A hardcover edition of this book was published in 1999 by Avon Books, Inc.

DEEP TIME. Copyright © 1999 by Abbenford Associates. All rights reserved. Printed in the United States of America. No part of this book may be used or reproduced in any manner whatsoever without written permission except in the case of brief quotations embodied in critical articles and reviews. For information address HarperCollins Publishers Inc., 10 East 53rd Street, New York, NY 10022.

HarperCollins books may be purchased for educational, business, or sales promotional use. For information please write: Special Markets Department, HarperCollins Publishers Inc., 10 East 53rd Street, New York, NY 10022.

First Perennial edition published 2000.

Designed by Kellan Peck

The Library of Congress has catalogued the hardcover edition as follows:
Benford, Gregory.
    Deep time : how humanity communicates across millennia / Gregory Benford
    p. cm.
    ISBN 0-380-97537-8
    1. Environmental risk assessment—United States. 2. Radioactive waste sites—Environmental aspects—United States. 3. Waste Isolation Pilot Plant (N.M.). I. Title.
GE150.B46   1999   99-41085
333.7'14'0973—dc21   CIP

ISBN 0-380-79346-6 (pbk.)

00 01 02 03 04 ❖/RRD 10 9 8 7 6 5 4 3 2 1

To
Mark Gregory Benford
voyager into still
deeper times

. .
.

CONTENTS

· ·

DEEP TIME

..

INTRODUCTION

From Here to Eternity

But at my back I always hear
Time's winged chariot hurrying near
—ANDREW MARVELL

Modern technology projects our grasp across great distances. In half a century our mastery has extended from the thin skin of our air to the rim of the solar system. Far less obvious is our growing reach through time.

Looking backward is a major intellectual industry. Archeology's ever-sharpening tools can now peek at the earliest towns, into the burial pits of the neolithic and beyond, to our simplest tools. Biology has read fragments of DNA from a fossilized magnolia leaf about twenty million years old. Evolutionary theory speaks almost offhandedly of great sweeps, epochs piled upon epochs, until the fates of mere species are as dust in the wind.

Astronomers see even further. We are used to thinking of vast epochs as seen through our telescopic time machines. We study the ruby splendors of galaxies as they were before the Earth was born, because light's finite speed has kept these images in storage as it propagated to us. With this so-called "look-back" time, we can witness billions of years into the past. This culminated in the recent discovery of small variations in radiation from the early universe—light that has so cooled by the later universal expansion that its temperature is less than three degrees Absolute. These tiny bumps we see across the sky, using microwave antennas, reveal structure as it was about fifteen billion years ago, only a few hundred thousand years after the universe began.

Modern tools peer backward through tree ring dating (good to about five thousand years), Arctic ice bores (which measures layers of ice, and the air trapped in them, good to about 200,000 years), and nuclear dating methods (good for the Earth's age, about 4.6 billion years). A variety of other methods discern ages along this scale with ever-growing resolution.

We are much less aware of our reach into the future. The *Voyager* probes of the 1970s carried plaques rhapsodizing over our culture, gestures at eternity cast into the interstellar abyss, possibly to be read billions of years from now. The high vacuum of space preserves well, but at the price of putting any message beyond ready human grasp.

Conversely, leaving messages close at hand runs great risks of obliteration. On Earth, until this century, the Pharaohs were the champions at knowingly reaching down to their posterity, a compass of less than six thousand years.

Our twentieth century cultures can do far better. This book will focus on how our species has, in a single century, begun to have effects that shall ring down through many millennia in unprecedented ways. Based on my own experiences, it moves from concrete cases to abstract proposals. My agenda is to encourage a deep time outlook, arguing that it may help us with our most vexing problems as a civilization.

Increasingly, such perspectives emerge from the reach of our technology. Rockets can carry indelible plaques to the stars, and nuclear waste can demand that we mark sites for times longer than the age of our civilization. More subtle impacts shall come from our inadvertent effects upon our planet itself.

Many works about archeology, astronomy, and evolution portray great stretches of time, but seldom do they confront our growing influence on events millennia hence. This book treats several examples of the restless energies that seek to reach across the ages.

Through millennia, we have tried to leave some testament to ourselves. Some are unconscious, others are nonverbal, and some are accidental or unintended.

Why? Surely the global character of this impulse tells us much about ourselves, in ways that the customary chronicles of history do not.

Monuments meant to last seem a largely male interest. Of course men have ruled for nearly all of history and so had the resources to build great works. Women were often illiterate and powerless.

There are few lasting records of women, though some wonders have celebrated them: the Taj Mahal is a mausoleum for a shah's beloved wife, and the Parthenon once housed a statue to Athena. We might glibly guess that women achieve a sense of perpetuity through children, and men labor to secure something similar. Still, the impulse seems quite broad.

I suspect that deep within us lies a need for continuity of the human enterprise, perhaps to offset our own mortality. This explains why so many deep time messages and monuments have religious elements or undertones. A yearning for connection also explains why ancestor worship appears in so many cultures; one enters into a sense of progression, expecting to be included eventually in the company. Somehow in the human psyche a longing for perpetuity has manifested itself powerfully, erecting vast edifices and burying considerable treasure—all to extend across time some lasting shadow of the present.

THE TEETH OF TIME

Within the last two centuries our appreciation of the expanses of time, fore and aft of our own precious Now, has expanded enormously. Two centuries ago Schliemann had not yet unearthed Troy, and Napoleon's forces were so oblivious of the importance of antiquity that they supposedly shot the nose off the Sphinx for target practice. (Recent study suggests that vandals removed the nose by hand, though.) French Pleistocene cave art was defaced in the 1800s with signatures (thoughtfully dated), in part because the visitors had no idea of their vast age.

Advances in radioactive dating and astronomical cosmology have left us standing, as a species, on a vast plain, perspectives of time stretching from our murky origins to the universe's cosmological destiny. This is a recent condition, quite modern. Ancient societies assumed a comforting stasis, that life and culture would go on for long, essentially infinite eras, sharing a common perspective and even religion. Whipsawed by incessant, accelerating change, the modern mind lives in a fundamental anxiety about the passing of all referents, the loss of meaning.

On the scale of a mere century we individually die. To persist beyond this means to survive through surrogates: family; nation;

schools of thought in philosophy, science, or art; religious communities. We have evolved with passionate loyalties to these larger units, probably because they do promise continuity, a consolation for personal mortality.

Over a millennium, neither politics nor technology are sure standards. Only languages, religions, and cultures retain their identity. A thousand years ago Europeans were crude villagers on the edge of the advanced civilization, the Arabs; but the seeds of Western emergence lay in their culture. Over such spans, only a strategy of what I shall call deep time messages can suffice to propagate anything—an idea, remembrance of a person, cultural works, or even a simple signature.

So far, ten thousand years is the upper limit of conscious, planned deep time communication. Not coincidentally, this is roughly the age of civilization. Little comes to us from beyond this scale except crude signs, notches in stone or antlers, mute stacks of stones, and cave paintings of mysterious intent. Ten millennia ago we lived in hunter-gatherer tribes just hearing about a hot new high-tech approach: agriculture.

Our numbers then took off under that most important of all technological revolutions. Agriculture and fishing appear to have been driven by necessity, as our burgeoning population made old hunter-gatherer modes inadequate. The efficiency of planting seeds and harvesting in turn benefited from the warmer climate coming after the ending of the last ice age. Soon came cities, many novelties, and enough amassed wealth to build more permanent, stony tributes to the powers of the day. Quite quickly the Egyptians and Chinese began erecting monuments to themselves. The impulse seems buried deep within us.

Such early testaments convey pride, even grandeur, but little more. Many ancient monuments are unmarked and mysterious, like the Sphinx, Stonehenge, and the American mounds. Probably most were not tributes to their builders, but religious sites or mausoleums. Deeper motives may have pervaded societies, which we, at our great remove, can only dimly sense.

A puzzle of far antiquity is why the ancients often built with great stones, moving burdens intimidating even to modern engineers. Managing a hundred-ton rock is far more difficult than placing ten ten-ton stones. Yet scattered over the lands of ancient civilizations are countless large stoneworks. At Ba'albek in Lebanon an 800-ton boulder still stands, carefully placed to form a temple wall. The temple's monolithic columns are equally massive. Such feats give clear evidence that the ancients could build on scales comparable to ours, through hard, protracted labor.

Such sites provoke awe, and the sheer numbers of large stone-works argues that techniques for building them were broadly known and highly developed. Some archeologists, seemingly innocent of engineering finesse, invoke the "more guys with ropes" explanation to explain how such works came about. More likely, specialized equipment and perhaps traveling artisans helped.

Recently an engineer charged with erecting a monolith of Stonehenge scale devised a counterweight method to tip the forty-ton stone into its support hole. Laying a wooden rail atop the horizontal stone, he put a heavy rock on the rail, near the larger stone's center. A small team then pushed the rock weight to the end of the monolith, levering it up until it slid into the slot, standing tall. Probably such tricks made ancient works far easier than the "ramps, ropes, and sweat" style often assumed. Further, such feats could give a sense of control over daunting masses, which may have been an enduring satisfaction for the entire society. Look what *we* did, such works proclaimed to generations unborn.

These surmises about ancient motivations seem plausible, but we must remember that they are guesses made through our cultural filters. Some societies (China, Latin America) think in terms of family dynasties, making investments that bear fruit fifty or a hundred years downstream, and passing on homesteads. Ninety-year mortgages are not unknown.

In contrast, our modern attention span is usually quite short. Most industrial societies have an increasingly bottom-line attitude. Stocks had better show a good quarterly statement, and long range research is uncommon in industry. In this century, many countries have failed to outlive their citizens. Physicist Hal Lewis wryly notes that "there wouldn't be so many proverbs exhorting us to prepare for the future if it weren't so unnatural." Most people consider their own grandchildren the furthest time horizon worth worrying about.

Such views are quite sensible. Why invest thought and effort in such chancy pursuits? Over ten millennia, qualitative changes dominate quantitative ones. Even fervently held values and ideals are totally plastic. Tempocentric notions of "the human condition" do

not survive. Confronted with one of our current skyscraper monoliths of glass and steel, what would a citizen of the year 5000 B.C.
think? No doubt these soaring towers would provoke awe. On the
other hand, what perspective would a person of the year A.D. 5000
bring? That ours was a great era, perhaps—or merely that for some
reason, possibly without noticing, we made our grandest buildings
in the same shape as our gravestones?

Indeed, our current concern for the past may not be itself long-
lasting. We moderns have watches and clocks to fix us in the immediate moment, ticking off each second. Some of our notorious
anxiety probably stems from these ever-present reminders. Paradoxically, we have leisure and inclination to study the past as never
before. Both these aspects may change.

Dire circumstances—and nearly all history can be described so,
compared with our luxurious present—shorten people's interests
and attention spans. In our era, high culture has increasingly
reached backward in time, expending great efforts in archeology
and other sciences, almost as if we seek our identity in distant ancestors.

The low culture form of this is nostalgia, and as cultural critic
Dean MacCannell notes, nostalgia may come from our notion of
progress:

> The progress of modernity . . . depends on its very sense
> of instability and inauthenticity. For moderns, reality
> and authenticity are thought to be elsewhere: in other
> historical periods and other cultures, in purer, simpler
> lifestyles. In other words, the concern of moderns for
> "naturalness", their nostalgia and their search for au-
> thenticity, are not merely casual and somewhat deca-
> dent, though harmless, attachments to the souvenirs of
> destroyed cultures and dead epochs. They are also com-
> ponents of the conquering spirit of modernity—the
> grounds of its unifying consciousness.

Associating the past with naturalness is often unconscious, and
we shall meet this idea again.

Time itself isn't what it used to be. We moderns labor under a
sense of linear time that emerged forcefully after Pope Gregory XIII
imposed the Julian calendar on the Catholic world in 1582. Linear

calendars had been around from the ancient world, but drifted out of synchronization with the seasons because of bad fits to Earth's actual orbital period.

Astronomical measures of duration embody only one of several concepts of time. Social time might be defined as the cycle of events according to beliefs and customs, subject to language and even fashion. Cultures can conceive time and space less abstractly, as in traditional Chinese concepts, which held that time proceeds by felt cycles, as mirrored in weather and sky. They imagined time to be "round," whereas space was "square."

Further, media reflect emphasis on either time or space. Heavy materials such as thick parchment, clay, and stone stress time and endurance. Media emphasizing space-saving are apt to be light and less durable, such as papyrus and paper. These are suited to easy dispersal of information and are prized by administrations, which have short attention spans. We hear down the corridors of history from either the original, durable media, or the flimsy forms that must be continuously renewed, as in the copying of ancient texts by monks in medieval times. Our century's electromagnetic media, from radio to the optical disk, are more perishable still.

In a sense, all technologies are attempts to contest the ordinations of time. Agriculture tries to make crops grow to order, medicine delays the onslaughts of age and death, transportation moves us faster, communication media strive for speed and preservation of information. There is a touch of eternity in the photograph, a technology for preserving the moment that would have astonished the ancients.

Beginning in the nineteenth century, critics sought to undermine the very notion of timelessness. They held that monuments mediate memory and insist that remembrance remains inert, moored in the landscape, ignoring the essential mutability of all cultural works.

Nietzsche disdained any vision of history that pretended to permanence. Lewis Mumford pronounced "monumentalism" dead since it clashed with his sense of the fluidity of the modern. "If it is a monument, it is not modern, and if it is modern, it cannot be a monument." Lacking the quality of renewal, monuments gave "a false sense of continuity." He saw this as essentially a moral failing, since by not putting their faith in renewal, out of vanity the powerful then mummified the moment into a petrified immortality. "They write their boasts upon tombstones, they incorporate their deeds in obelisks; they place their hopes of remembrance in solid stones joined to other solid stones, dedicated to their subjects or their heirs forever, forgetful of the fact that stones that are deserted by the living are even more helpless than life that remains unprotected and preserved by stones."

Even quite recently, some find memorials destructive. Pierre Nora warns, "Memory has been wholly absorbed by its meticulous reconstruction." As James Young remarks, "To the extent that we encourage monuments to do our memory-work for us, we become that much more forgetful."

These views stem from short horizons. "Memory-work" necessarily transforms and ebbs as centuries roll on. Legends warp. To be sure, in broad outline, folk memory is surprisingly long-lived. Modern Australian aborigines recall landmarks that were flooded since the last ice age, eight thousand years ago; divers verified their existence. But much of this information is cloudy; what does the mythical beast they call the "bunyip" correspond to?

The modernist fear of rigidity already seems a bit antique, since modernism has entertained newer ideas, including "postmodernism," which seeks to undermine the meaning of texts. (This seems a passing fashion, more a mistaking of momentary cultural exhaustion for a fresh, innovative view.) It seems likely that antimonumental thinking is fading faster than will messages that attempt to speak across gaps of language, culture, and intention.

Our own individual pasts get filtered by later experiences of time's flow. It is commonplace to note that the years flicker by faster as we age. Certainly a new year can have less impact when we have many more stacked behind us. I suspect the sameness of the later years also alters our reading of them. We settle into habits, and the days have fewer distinctions to mark their passing. We slide forward on skids greased by routine.

Little wonder, then, that we have a keener sense of the endless centuries behind us as our expected lifetimes approach a century. To a baby, a year is like a lifetime because it *is* his lifetime, so far. By age ten, clocks tick on at an apparent rate ten times faster than the baby's sense; the next year is only a ten percent increase in his store of years. At fifty, time ticks on five times faster still. At a hundred, the differential rate is a hundred times the baby's.

Some poets have found this a blessing, as in Thomas Campbell's "The River of Life":

> *Heaven gives our years of fading strength*
> *Indemnifying fleetness;*
> *And those of youth, a seeming length,*
> *Proportion'd to their sweetness.*

Imagine living to a thousand; then a year would have the impact of a few hours in a baby's life. To such a being, deep time is the proper scale. In later portions of this book I shall argue that this perspective is one we should entertain as a society, as our powers grow and our problems confound us.

In thinking of far antiquity, we cannot help but invoke our current assumptions. In the 1990s, historical analysis often assesses our past using current moral or ethical standards, a critical posture doomed to obsolescence as tastes change. Something broader and less bound up in the moment is needed. This book is a contribution to such perspectives.

Culture shapes our vision of the past, even grossly falsifying it. As well, memory is notoriously unreliable. Individual recollections of the past are easily and quickly shaped by others and after a while need have little bearing on the once-lived events. Consider how many believe one or more of the conspiracy theories about the Kennedy assassination.

Deep time messages seek to counter this, consciously or not. We are often unaware of how antiquity influences us, for as we shall see, some signals across the abyss of deep time we do not even recognize as artificial.

Throughout history, most people—as opposed to some institutions—have never given thought to the morrow beyond their own grandchildren. We moderns have taken this to new heights. Yet attempts to affect distant generations appeared in early civilizations. As we shall see, we live in a world subtly altered by changes wrought before historical recording began.

These themes I shall explore in the specific ways they entered into my own work as a practicing physicist and consultant. My intuition is that by grasping specifics, we can sense the dimensions of this shadowy subject.

LEAVING TESTAMENTS

*It is easy enough to say that man is immortal simply because he will
endure: that when the last ding-dong of doom has clanged and faded
from the last worthless rock hanging tideless in the last red and dying
evening, that even then there will still be one more sound: that of his
puny inexhaustible voice, still talking.*

—WILLIAM FAULKNER
NOBEL PRIZE ADDRESS, 1950

Assurbanipal, king of Babylonia, Assyria, and Egypt in the seventh
century B.C., amassed a vast library of stone tablets laboriously
incised with the knowledge of the day. Today these comprise a
useful trove for scholars. Assurbanipal was following the lead of his
father, Esarhaddon, who buried cuneiform inscriptions in the foundation stones of monuments and buildings.

They obeyed an impulse common to virtually all cultures. Typically, the practice springs from a class that feels it has accomplished
much and has the resources to leave durable messages announcing
this. The universality of this impulse is fundamentally positive and
far-seeing, time-binding us with generations before and after our
brief moment in the sun. Practiced over millennia but seldom noticed in the everyday rhythms of our lives, the desire to pass on
messages gives us perspectives on the import of our own actions,
seen against the long odyssey of our species.

There seems to us something fitting, elegant, and deeply human
in such gestures reaching across the abyss of time, a humbling ack-
nowledgment that posterity is quite real and important to us. Yet
such acceptance is oddly exalting, too.

Such sentiments readily emerge from contact with ancient monuments. More complex and ambiguous feelings come in the face
of the oldest concerted attempts to leave creative records, the cave
markings found principally in Europe.

Were the cave painters hoping to send some record of themselves
down through deep time? As usual, we can only speculate; paintings seldom announce their intentions. Many have sensed that the
cave art did contain messages, but increasingly, after decades of

warring theories, experts believe that we cannot understand the messages clearly because they are not aimed at us.

Most commonly, anthropologists believe the paintings had some magical purpose. Did showing spears or harpoons penetrating game ensure a good hunt? But such weapons appear seldom. There are even counterexamples, such as a scene from the "Dead Man's Shaft" in the famous Lascaux cave. A realistically pictured bison is goring a man, who is childishly drawn. The bison is also wounded, impaled by a spear, its intestines protruding. Was this detail considered important enough to chronicle with care? Then why is the man crudely done?

Others believe that the paintings are art for art's sake, period. Since some anthropologists believe these people had plenty to eat and leisure time, this seems plausible. Though the work ranges from bare, artless graffiti to stunning depictions, they all share a precision of observation. These artists knew animal behavior and fauna down to small details, and rendered them exquisitely.

This suggests that many paintings may have aided instruction of the young. Gathered safely inside, by fireglow young boys and girls could learn how animals gave away their movements and moods and methods. Some paintings begin near cave entrances and then fade toward the sunlight, erased by time, suggesting that they continued outside. Given the ease and pleasures of working outside, we can guess that ice age humanity may have left innumerable works on rocks, trees, and boulders, of which a tiny fraction have come down to us.

Crucially, we cannot know if they had any sense of long time scales, or any urge to leave their mark for the shadowy far future. But the impact of their message, whether intended for their children or as art for art's sake, shines through. These very ambiguities make us study their works all the more.

Perhaps our biases in reading such ambiguities come only after civilization arises, with its firm buildings and records. The ancient Greeks, surely the most influential culture of all time, encapsulated this by making lists of the monumental constructions they found most awesome, labeling them the Seven Wonders. Only one of these Seven Wonders of the ancient world stands today, the Great (Cheops) Pyramid. In a sense, all the Seven Wonders were messages intended to provoke in us remembrance mingled with awe, and as such, six have failed.

A tour of the sites of the Seven Wonders is instructive. No ancient could have seen them all, since they did not all exist simultaneously. There were several ancient lists of the Seven Wonders, each heavily favoring the Greeks, who wrote all of them. The Palace of Cyrus, king of rival Persia, was discreetly ignored. The Temple of Solomon in Jerusalem might have made the list, but was a shrine to a barbarian god, after all.

The fabled Hanging Gardens of Babylon, of which we have not a single authentic image, seems to have been abandoned and scavenged within centuries. No trace remains of them, and we have only a vague idea (from texts) of how the hidden plumbing kept them lushly green.

In present day Turkey stood the Temple of Artemis, built at Ephesus in 550 B.C., when the region was in the Greek cultural orbit. Writing in A.D. 60, Pliny the Elder called it "the most wonderful monument of Grecian magnificence." Made of white marble, it was 425 feet long, 225 feet wide, and with its more than one hundred 60-foot stone columns, supported a massive roof. Its ornamental sculptures and paintings were said to be of extraordinary beauty. Today a lone bare column sticks up from a muddy field, so unremarkable and unmarked that visitors often drive right past the site. Burned down in the fourth century B.C., the temple was rebuilt in the third and then sacked and destroyed by the Goths in A.D. 362.

The most renowned sculptor of antiquity, Phidias, created two of the Wonders: the Colossus of Rhodes and the Olympian Statue of Zeus, around 435 B.C. The bronze Colossus stood as tall as the Statue of Liberty and was even more massive, yet fell in an earthquake only fifty-six years after it was erected. They were both vandalized by invaders within centuries. No authenticated parts survive.

The Mediterranean was a pleasant, warm region for developing human culture, but its crust was unstable. Several of the Wonders were damaged or destroyed by earthquakes. The Lighthouse of Alexandria, ca. 280 B.C., stood a striking 350 feet (105 meters) high, dominating the harbor, but fell in an earthquake in the thirteenth century.

Of particular poignancy was the Mausoleum of Halicarnassus (ca. 352 B.C.) on what is now the Turkish coast. The widow of the tyrant Mausolus finished this tomb, spending a fortune embellishing the tower with the finest statuary and encrustations of the Mediterranean basin. Mausolus apparently began his own tomb construction, echoing the Pharaohs, who had work start upon their ascension to the throne; ruler cults demand constant attention.

As described by Pliny in the first century A.D., this magnificent tomb struck his era as comparable in impact with the pyramids. Tapered using three step-backs, the tomb rose some fifty meters to an artistic climax: thirty-six columns in colonnade supporting a roof whose twenty-four steps carry a colossal four-horse chariot of marble. With elaborate friezes carved in relief—one of battle scenes with centaurs, another with Amazons—and sculptures ranging from natural to colossal in size, its point lay below, where a staircase nine meters wide led down to the tomb chamber. Mausolus was cremated before burial on a huge public pyre. His tomb was crammed with riches (all stolen later), stocked with slaughtered calves, cows, sheep, and hens, and cleverly drained by ducts along its walls. A solid pile of stone blocks filled the staircase, was covered with earth, and finally a plug rock barred entrance. Robbers plundered it by digging a tunnel into the rock beneath the foundations.

Mausolus ruled as a Persian satrap, filling his treasury by taxing even those who wore their hair long. He was shrewd, unscrupulous, and much feared and hated by his subjects, who several times attempted assassination.

The tomb gained its grandeur not from scale but from ornamentation, and this may well have proved its undoing. Shaken by a distant quake, statuary slid down the roof and crashed into other figures, such as the large stone lions that stood guard in front. There is archeological evidence that this occurred. Shifts of the rock made the marble facing crack and peel away. Such degradations apparently disfigured the building long before it fell. Finally, the structure proved top-heavy and vulnerable, so that a final quake in A.D. 1304 brought it all down. The Crusaders investigated the ruin, dug and found the burial chamber, but were forced by the raids of Arab infiltrators to abandon it at nightfall. When they returned the next day, the tomb had been looted, and much of the testament to the greatness of the tyrant was gone. It was never recovered. Later these same knights used the ruins as a quarry to erect a castle that still stands today.

When I visited the site, littered with drums of stone that once were part of the grand columns, there remained only an air of shattered grandeur. Like several of the Wonders, this monument could have been repaired after quake damage, but the urge to do so had ebbed away. Mausolus's primary legacy is a word, mausoleum.

Not even all of the pyramids of Egypt survived. The vast mudbrick pyramids sustained serious quake damage. The large pyramid at Medium is today a three-step pygmy compared with the original, whose limestone outer casings form a jumbled skirt around it.

The Pharaohs apparently built the pyramids to solidify their hold on the world's first and greatest thantocracy. Organizing society around death demanded convincing demonstrations of mastery by those who said they held open the portal to eternity. Shaped to call forth comparison with mountains, the pyramids were the largest objects visible within the narrow world of the river-centered nation. They made Pharaonic power obvious to all and promised a firm solution to the most basic of all human problems, the dilemma of death. Construction provided worthy labor to the peasant masses during the Nile's flood season, when idle hands might make trouble for the state.

The pyramid pinnacles carried capstones sheathed in gold; the only surviving specimen is covered in religious symbols. Greek travelers beheld the Great Cheops Pyramid clad in fine white limestone, a dazzling sight from many miles away with its gold peak. The limestone was stripped away millennia ago to build Cairo. We see only the core, rugged and massive and still awe-inspiring. Indeed, the pyramids were man-made mountains, proclaiming that the state could echo nature's feats.

Critics have berated the Pharoahs for spending great wealth on such useless monuments. Of course, the Pharoahs did not see them as useless, but rather as a sure way to gain a pleasant afterlife. Further, Egyptian engineers learned much about quarrying, shaping, and maneuvering large blocks, learning that passed into humanity's general knowledge.

The Egyptians set a pattern seen often in antiquity. Ancient palaces were mostly built of mud brick, and eroded quickly. Since this passing vale of tears was far less important than an eternal afterlife, this was proper, just as building tombs and temples of stone fitted their enduring importance.

A funereal air surrounds lasting monuments down into our time. Tombs on aging European estates have outlasted several great houses nearby. Of course, even mass does not guarantee that your message will remain in context. The largest obelisk of ancient Egypt was 105 feet long and now resides in Rome, its intention ignored.

We cultural descendants of the Greeks have canonized the Seven Wonders, but equally powerful works appeared in antiquity, far from the Mediterranean basin. The Great Wall of China, the only ancient construction so large it is visible from space, was unknown to classical Western civilizations. The huge dam at Ma'rib in Arabia and the Buddhist stupas of Ceylon surely would have made the list if the Greeks had voyaged farther. The Nazca Plain lines and Easter Island heads, while more recent, might have made the list.

Perhaps the best known architecture of antiquity is the Acropolis of Athens, with its Parthenon and other buildings. It was not on most of the several Seven Wonders lists. Dating from the fifth century B.C., it was a transcultural religious site for Greeks, Romans, Christians, and Muslims. Some metal statues were "recycled" in antiquity. Removal of some pieces to indoor museums (notably the Elgin Marbles to London) prolonged the durability of some statues, friezes, and columns and spread the Acropolis's renown as a cultural artifact worth preserving. Only modern acid rain has damaged it appreciably, arguing against returning the scattered pieces to the original, outdoor structures. The crowning glory of the Acropolis, the Parthenon, has become a pervasive standard for architectural beauty.

Strikingly, no libraries survived antiquity, though some were quite grand. A Christian mob burned the greatest trove of ancient writings, the Library of Alexandria, taking from us hundreds of thousands of papyrus and vellum scrolls. Writing on organic sheets is vulnerable to fire, whether from fanatics or accident. Acid-free paper withers in a few centuries.

Stone lasts. It is still the wisest deep time investment.

The oldest reliably dated structure in North America is a 5,400-year-old earthen mound at Watson Brake, Louisiana, fully two thousand years older than the much better known, classic mound-builder sites of other river valleys. Thousands of artificial mounds dot the USA's East and Midwest, shaped like serpents, giant cones, or square platforms. Though some were used as ceremonial centers and slaughterhouses, their purpose remains mostly mysterious.

The civilization that built them had no writing or pictorial displays, yet sent the simple message of impressive large structures across millennia. Their prolonged earnestness is clear, since the Watson Brake site apparently took four hundred years to finish, yet trade and agriculture do not seem to be the primary motivations behind the builders. Asked for their purpose, an archeologist remarked, "I know it sounds awfully Zenlike, but maybe the answer is that building them was the purpose."

No one ever lived in this mound area. Perhaps it had a special aim beyond the everyday. Such long-lived sites transmit a blunt signal of existence, no more. Perhaps this makes them all the more compelling, for the silences of such sites seems to have a significance of its own.

This brief tour of antiquity suggests that the high culture of many civilizations sought to propagate or commemorate itself in permanent ways. That they often failed only underlines how difficult the task is. We know of many failures, but can only contemplate the probably larger number we do not know. Doubtless, many went to their graves believing some sliver of their identity would ring down through the corridors of time. Few did.

FUTURE WONDERS, HIGH CHURCH AND KILROY

In 1995 the American Society of Civil Engineers produced a list of the Seven Wonders of the United States. Two were holdovers from a similar list they compiled in 1955, the Hoover Dam and the Panama Canal. Dropped from that earlier list were the San Francisco Bay Bridge, the Empire State Building, the Grand Coulee Dam, the Colorado River Aqueduct, and the Chicago sewage disposal system (surely a wonder suited particularly to engineers' discernment).

All these still proudly stand, but apparently the engineers now find them less wondrous. In merely forty years they have yielded to the Golden Gate Bridge, Manhattan's World Trade Center, the Kennedy Space Center, the Trans-Alaska Pipeline, and the overall Interstate Highway System. While the first two provoke awe, the others do not possess the singular impact I associate with wonder. This suggests that our sense of the awesome is quite personal. Deep time messages must speak to feelings of wonder that persist across both time and culture.

Arthur C. Clarke mentioned to me his choice of the Seven Wonders of the Modern World: the Saturn V rocket, the microchip, the rock fortress of Sigirya (a Sri Lankan temple invoked in his novel, *The Fountains of Paradise*), the Mandelbrot Set (a mathematical figure which set off the vogue for fractals), Bach's Toccata and Fugue in D Minor, the giant squid, and SS-433 (an astronomical object that projects twin jets of matter in opposite directions from its core, moving at a precise fraction of the speed of light).

This is really a mixed list of accomplishments and objects, the last two natural; presumably he means that discovering them is a wonder. Clarke's list betrays a distortion familiar in ancient inscriptions: up close, events loom large. Minor battles seem to rival Gettysburg. Stature dwindles along with immediacy. His addition of the microchip suggests a difference between our present technological triumphs and those of the past. We can be inspired by quite tiny things which pack enormous information densities into their compass. But will such passing marvels provoke wonder in future minds?

This question suggests that we study the primary deep time communication methods used in the past, hoping they will work into the future.

Undertakings to convey the best of current culture seem to have had the best chance of surviving, for their beauty could protect them for at least a time from random vandalism. These attempts I shall call *High Church*: communicating the culture of the upper crust, usually solely of the politically powerful class. Often the High Church strategy conveys more than it intends; for example, we read into Egyptian hieroglyphics class and economic issues that the artists assumed natural and unremarkable, hence invisible.

A parallel note sounds down through the millennia: *Kilroy Was Here*. This rather mysterious graffito emerged in the twentieth century, the point being perhaps that nobody knows who Kilroy was, but he (she?) thought himself important enough to leave his mark. Indeed, who among us does not?

Kilroy is an old phenomenon. The last of the major step pyramids of Egypt is at Saqqara. It set the trend in design, today appearing as a high rectangular tower with a base engulfed in sand. A thousand years after it rose, someone scribbled on a wall of its mortuary temple that he had come "to see the beautiful temple of King Snefru," though we are not sure exactly who was buried there. The sight was glorious: "May heaven rain with fresh myrrh, may it drip with incense upon the roof of the temple. . . ." Other graffiti festoon the pyramid. Greek mercenaries of 700 B.C. left their names on many Egyptian monuments.

Indeed, even the famous are not immune; Lord Byron's name is carved into the Temple of Poseidon at Cape Sounion. In the early 1800s a British traveler in Egypt incongruously named Giovanni Belzoni flamboyantly cut his appellation into many monuments, making himself well-known to modern visitors by sheer irritating persistence. Though now disparaged by Egyptologists, he accomplished the aim of Kilroyism: sending at least his name across the abyss of years.

This simple, universal impulse gives us the most common type of deep time message. In nooks and crannies of once great buildings, one finds the cut or carved or scrawled evidence of a desire to not go unheralded into oblivion: *Kilroy lives*.

MEANING WITHOUT WORDS

We read messages everywhere, even when there aren't any.

Everywhere, cultures perceive constellations and stories in the stars. We see faces where there are none, in the chance roiling of clouds, the dark lava features of our moon, or the erosion patterns of a mountain on Mars.

We also read buildings. No architecture stands independent of aesthetics. "Beauty" changes with time, so that buildings once considered masterful are often demoted and even demolished. The High Church defense, then, depends on how long one thinks a given cultural preference shall reign. Future Kilroys lie in wait. As urban planners know, the first signs that a neighborhood is beginning to slide are graffiti.

Those considered beautiful can even ignore structural rules. The Athenians converted a wooden temple into a stone one, disregarding the vastly different demands of the transition, yet creating the marvel of the Parthenon.

Architecture conveys many nonverbal, or *semiotic*, messages: consider the barred windows of a jail versus the ornamented, status-rich windows of a Renaissance palace. Semiotic meanings are deeply cultural and time shapes them. The pyramids once sent a religious message, later a magical one, still later an engineering one; today we see them mostly through a lens that combines artistic, structural, and sociological ideas. Still, they have great power. Doomed nuclear-war survivors in Nevil Shute's *On the Beach* spend their last days constructing one in the Australian desert, a monument with no future audience, since humanity is dying; yet they build.

But some themes seem embedded even more deeply in the way we see our world. From the tension in a rope pulling a bucket from a well we intuit tensile properties. These then inform our way of looking at a suspension bridge, surely the most elegant combination of function and form in our time. Semiotic messages satisfy us best when we understand them both structurally and aesthetically.

A sense of wrongness can also be deeply intuitive. A tree trunk and branches tells us, by analogy, about gravity loads in tall buildings, adding from top to bottom. Violating this intuition brings a sense of error or ugliness. We stand puzzled before Cretan columns, which thicken as they rise, but accept Doric columns that are widest at the base. Cantilever beams that thin toward the tip are right to us, while a beam broader at the tip strikes us as wrong. The Cheops pyramid reminds us awesomely of a mountain, but a building of inverted, truncated pyramid design says to us that some trick has been played to give an unnatural result. We get an uneasy surprise from glimpsing one on the horizon, of "dishonest" design.

Scale does not seem to impede semiotic messages. Spiderwebs and the Golden Gate Bridge alike call forth our perception of tensile structure, their lightness striking us as obviously elegant.

Yet semiotic readings can change quickly. The Eiffel Tower was vigorously opposed and originally slated to be torn down after the exhibition it ornamented was over. Within a generation this "monstrosity" had become the very symbol of Paris and France itself. It attained this glory by conceding almost nothing to decoration, revealing its sinews completely, like a vertical suspension bridge.

Perhaps the most reliable wordless message to send across the millennia is *awe*. To instill this mingling of fear and wonder leaves the visitor with a memory free of words or detail. Though this is a High Church approach, fear can be part of the effect: the convergence of awe with the awful. Many have seen something fleetingly terrible in the visage of the Sphinx.

Ancient things and places hold inherent wonder for us because they speak to deep aesthetic biases we share. The most striking of obvious markers hold a still and subtle balance that their makers carefully shaped, speaking across time in the language of beauty. Ancient cultures fell in line with nature, many of their most obvious markers—pyramids, astronomically aligned henges—comprising a vast, unvoiced aspiration to join in harmony with elemental forces. Their stones speak to us still.

Next to the Egyptian pyramids, Stonehenge seems the best known example of this. Somehow this ring of stones instills wonder, though at first glance it is not very impressive, to some. John Fowles in *The Enigma of Stonehenge* (1980) quotes a child saying worriedly, "Why are there so many doors?" To the untutored eye it seems easy to see in it a mere lot of doorways leading nowhere.

Like the several hundred "henges" in Britain, its true purpose is unknown. Plainly it meant much to the ancient laborers who dressed the huge stones, moved them many miles, pounded them level and true with mauls. Their sarsen stone is three times harder to work than granite, able to ruin the edge of most modern tools short of steel alloy. Though the site slopes several feet east to west, the tops of the lintels vary by only three inches, after millennia of settling. Engineers estimate that about half a million man-hours went into simply digging the ditches and banks of the area.

Explanations of its geometry revolve about astronomy; photos of it invariably portray the sun rising or setting between uprights and a lintel. To the modern mind, which rarely notes the rise of the sun nor of stars, the astronomical role is not intuitive. Millions today know their birth sign, but because the stars have moved since the Babylonians invented the Zodiac, most of them have it wrong. Such people may find it difficult to conceive of a passionate interest in getting the winter solstice, say, exactly right, as viewed through a ritual stonework.

Still, the henges seem to focus lines of sight and to shut out glare; they are more like sun visors than doorways. Even copies, such as the exact Stonehenge replica beside the Columbia River in Washington State (built, oddly, as a memorial to the World War I dead), capture this reverential flavor.

Stonehenge itself was an observatory, not in our modern sense of a place to discover the new, but rather, a site to embody knowledge of the sky won over centuries, if not millennia, and already old. The strength of conviction demanded to inspire such awesome labors, constructing an entire complex of henges (Avebury and Woodhenge lie nearby), all to reflect an understanding of our universe as revealed in the sky, is difficult for us to fathom. This underscores the immense cultural gulf that deep time messages must span, and so seldom do.

Even though awestruck, we do note geometrical messages, some no doubt unconsciously. Stonehenge's central stones form an oval as seen from above. The lengths of the two oval axes are in the ratio 5:3; this is close to the Golden Section, 1.6280 . . . , a number of great import to the ancient Greeks, and which we shall meet again. Such deep aesthetics can cross cultures.

It is also worth noting that the perimeter of the Great Pyramid, divided by its height, is 2π, so the height was set to equal the radius. The symbol of the sun god Ra was a circle, so when Ra rose at morning the pyramid greeted him with a geometric analog of himself, a hailing call from his subjects.

Such mathematical clues play to perceptions free of sentences. The Golden Section is a preferred number in the aesthetics of many different cultures. That these predilections appear to come down to us intuitively, while we must endlessly speculate on the true intentions of the Stonehenge builders, suggests that some very rarefied messages can persist.

Great monuments also seek to carry messages through the ancient language of mass. Bulk alone can draw our attention. Texts like the Bible carry messages through a hardening of an existing culture, protecting the text itself from tampering or extinction.

This desire to convey some essence of ourselves, whether High Church or Kilroy Was Here, is the great impulse behind deep time messages. But there is also a clear desire to shape the future, and to use the idea of the future to shape the present. Many legacies stem from this desire.

TIME CAPSULES

Love unto the uttermost generation is higher than love of one's
neighbor. What should be loved of man is that he is in transition.
—FRIEDRICH NIETZSCHE

The universal human urge to bury the dead, and often accompanying objects, may stem from the agricultural experience of burying a seed and seeing a specific plant grow later. But Neanderthals' careful grave burials belie this easy explanation, though quite possibly the impulse behind human burials is to invoke the resurrection we see in nature each spring.

In this sense some deep time messages play to this "natural" predisposition. The proposal for a Library of Life that I discuss in Part III invokes the storage and replanting of seeds, a Jungian archetype. But longtime marking of radioactive waste sites, described in Part I, is antiarchetypal, since we are planting not life but antilife, poison. Rather than say the "right" thing ("Take my husband, Crag, who lived a good life and deserves mercy"), our waste site markers must proclaim: "Stay away, danger!"

We have learned much from ancient burials. Intentional grave sites are often High Church, providing the deceased with some supplies for the afterlife, or artistically decorating casket, sarcophagus, crypt, or tomb. Inadvertent burials can tell us much more, as with the chance discovery of a man covered by ice for 5,300 years in the Austrian-Italian Alps. This unique find brought us a body well-preserved and carrying its microorganisms and parasites, his working tools (bows, arrows, dagger, axe) and clothes. Apparently caught in a sudden storm, his is the best permafrost mummy ever found, a trove of clues to a society and time that left no written records.

Bodies can even give us enough clues to reconstruct plausible features, literally giving faces to the past, as in a striking depiction of Alexander's father from an 8 B.C. Macedonian grave site. Science can pluck subtle clues from apparent ruin, making time capsules from accidents. Whole cities flooded by lava, such as Pompeii and Heraculaneum, provide us with encased bodies and buildings, invaluable archeological troves. As science advances, most deep time messages will be inadvertent, pulled out of history's noise level.

Time capsules embody a modern faith in a future that will care about us, underpinned by our anxiety that our most cherished beliefs and customs may be unintelligible, meaningless. Cities and institutions, no less than whole nations, have solemnly buried memorabilia, with much accompanying ceremony. The Order of Masons' cornerstone-laying ceremonies may have started this modern practice; in 1793 George Washington, a Mason, laid the original cornerstone of the U.S. Capitol, which may have held artifacts and has since been lost.

Roughly ten thousand time capsules already await future historians. Notably, their time horizon is quite short, usually a century. Citizens of Sandusky, Ohio, will presumably gather to open a capsule laid down only fifty years before, though filled with objects that recall "the triumphs and tragedies of life in America during 1995." These include Pop-Tarts, crayons, the May 29, 1995 edition of *Newsweek*, a "Buns of Steel" video, and a Wonderbra. In 2043, Euclid, Ohio, presumably will dig up a seven-foot torpedo tube packed with a history of town organizations including the Polka Hall of Fame and a "Not Too Young to Polka" cassette.

The Seville World's Fair of 1992 left an interesting democratic variation: the Capsula de Tiempo, an open tar pit into which anyone can throw whatever they wish. This cavalier style, advertised as "democratic," contrasts with the 1939 World's Fair tube of copper, chromium, and silver, "deemed capable of resisting the effects of time for five thousand years." It was the first to be called a time capsule, though its first name was "time bomb" since some believed its opening in A.D. 6936 would set off a cultural explosion, with its textiles and microfilm, TV set and a machine that teaches English.

For the U.S. Bicentennial, President Ford had 22 million citizen signatures collected for interment at Valley Forge, intending them for study in 2076. Proudly toured throughout the nation for display, the capsule was, however, stolen from a van at the burial site. This is another common theme: slips between cup and lip. Many capsules are already lost, their markers not erected, details of location forgotten. The city of Corona, California, has already laid down seventeen capsules (one of which I saw solemnly interred in 1963) over the last half century, only to lose track of them all, though

Figure I.1 The Crypt of Civilization at Oglethorpe University, Atlanta, Georgia, sealed in 1940, to be opened in 8113

they did tear up a lot of concrete around the civic center in a fruitless search. The cast of the "M★A★S★H" TV show buried a priceless set of tapes of the show, plus artifacts, somewhere in the 20th Century Fox parking lot in Hollywood, but nobody knows just where. Though buried only in 1983, it is already submerged beneath a huge Marriott hotel.

A 1953 two-ton capsule mandated by the state of Washington lies lost beneath the capitol grounds because, during political infighting, the legislature did not fund the last act, its marker. People can recall the ceremony, but not precisely where it was held.

The approach of the millennium has caused a boom in time capsules; there is even a company producing custom-engraved aluminum tubes. They seem designed to last at best a few centuries. Capsules are usually gestures of more importance to their planners than anyone else. The pop artist Andy Warhol filled 608 packing boxes between 1974 and 1987 with inexplicable memorabilia, ranging from unknown paintings to telephone messages, check stubs, and a piece of cake. He intended this trove for some permanent capsule, but never built it. In 1997, Pittsburgh's Warhol museum paid tribute to him by laying down another capsule assembled from whatever local residents brought in. As engineers put it, the signal-to-noise ratio here seems low.

Some capsule designers take the longer view. Oglethorpe University in Atlanta sealed a Crypt of Civilization in 1940 (see Figure I.1), not to be reopened until 8113. (This date is as far from 1940 as were the earliest dated writings then known.)

The crypt was devised by Thornwell Jacobs, who began the academic pursuit of time capsules with an article in *Scientific American* in 1936. He was impressed by a capsule buried by Tokyo citizens, to carry forward the names of ten thousand victims of a 1923 earthquake. Far greater disasters have befallen this century, so this tragedy and its Kilroy capsule now seem eclipsed as they lie in quartz jars beneath a Buddhist temple.

Oglethorpe houses the International Time Capsule Society, and its crypt is indeed vast, the size of a swimming pool. It contains microfilmed books "on every subject of importance known to mankind," artifacts dealing with six millennia of history, the inevitable photos and wire recordings of world leaders circa 1940 (already mostly forgotten), and a quart of Budweiser. The designer apparently felt that much would change, but not beer, though it seems an uncertain sort of Rosetta stone. Perhaps it will be reassuring to the openers in 8113 to find that this Bud's for them (though undoubtedly gone flat), but it is unclear how anyone will know when to open the repository itself.

This is a classic dilemma of deep time: safely buried, how does the object announce itself to its intended audience?

As the Vatican well knows, preserving a capsule, marker, or message by creating an attendant culture can work over millennia. Ancient religious sites such as Mecca transmit with high fidelity their central tenets, perhaps more concretely than their sacred texts, which can be Bowdlerized or reinterpreted, since they rely on perishable parchment or paper.

Still, no institution comes down to us intact from the vast era before the invention of writing. This is no accident; text carries so much information, it can knit together whole communities. Perhaps the success of the Catholic Church stems in part from its deep urge to copy old manuscripts, which served it well a millennium ago.

It seems unlikely that one could build such a devout community today, short of launching a new religion. In Osaka, Japan, a group plans to bury in 2001, atop Mount Fuji in Antarctica, a master time capsule housing biological samples; apparently they feel that isolation is the best defense.

Suggestions that nuclear waste sites be cared for by an "atomic priesthood" ignore the motivations of the priests. It seems difficult to imbue groups with the dedication to spread superstitions about spots having "bad juju." Skepticism suggests that some future Age of Enlightenment would spawn free-thinking types who would venture into the sites to prove that the priesthood was full of baloney.

Still, the rise of mass celebrations could conceivably lead to such a community. The current Burning Man assembly in the desert near the Nevada-California border, dedicated to torching a tall wooden statue every year, echoes the Celtic wicker man ritual but seems unlikely to convey any durable message, though cultural critic Stewart Brand has at least ventured the idea.

Brand and others have founded the Long Now Foundation, which seeks to leave an enduring clock and library, perhaps with a community to support it, probably in the Nevada-California desert. A clock marking off millennia could inspire long-term thinking, though how can anyone both make it public and protect it? With a nonsecular priesthood?

Brand cites an arresting fact: the oak beams in the College Hall of New College, Oxford, needed replacing in the nineteenth century, so the college cut down some oaks planted in 1386 for that express purpose. In our modern time-pressed lives, does any organization, even New College, plant or plan for such perspectives?

Bits can last better than mass; strong belief systems have weathered from antiquity, principally as religions or philosophies. The cohesion of the Jews is legendary. The Pharoahs' priesthood lasted millennia, and New College's faith in the continuity of English culture is striking to us in our helter-skelter times. We could ape such deep time strategies. But how to inspire a priesthood? Modern times have been full of convictions, many strongly felt (Communism, Fascism, Socialism) and short of span—but it is not an age of faith.

At the end of our millennium we face a particular problem: the rate of change drives short-term thinking, but as our powers increase, our problems become longer-term. Environmental impacts are the best example. Meanwhile, our principal tool, technology, is moving toward the transient and small.

Our modern sense of the technological sublime stems from fresh sensations of size, speed, sound, and novelty. Big things compel attention, as always; even better if they are loud and fast and new. The Apollo V rocket fits all of these.

Our newest technological marvels are relentlessly small and quiet, though, from ever more compact computers to genetic engineering feats. When Arthur C. Clarke picked the microchip as a recent wonder, his response was intellectual, not visceral. Speed, compactness, and novelty are passing wonders. Again, stone is the best deep time investment.

Unlike the Hoover Dam, which was designed to last two thousand years (and has impressive symbolic star-chart time markers that could be read over that time), the late twentieth century leaves few impressive technomarvels. The High Aswan Dam and China's Yellow River Dam carry no notable messages. The millennial fever surrounding 2001 comes in an age more time-obsessed than any, but whose latest technology seems particularly inappropriate for deep time messages.

Still, one could bury a time capsule and hope for the best. Once only kings could marshal durable deep time messages. Now we all can at least try.

THE LONG FUTURE

If we are to survive through a long future, we must stay in contact with our long past.

—FREEMAN DYSON

In his *Imagined Worlds*, Freeman Dyson uses Shakespeare's seven ages of man from *As You Like It* to outline a grand perspective of our possible futures. He sets the seven ages as "not the seven parts of an individual life but the different time-scales on which our species has adapted to the demands of nature."

These scales are handily written in powers of ten: 10^x, where x runs from one to seven. At the upper end, ten million years, the major primates evolved. Similarly, the best any deep time message can envision is communication to the last members of our species. This takes us to the scale of x between 6 and 7, when evolution may well find a different shape and portent for intelligence. For messages to survive beyond a million years demands that we place them beyond the reach of human intrusions and the rub of wind and water. This means launching them into the preserving vacuum of deep space. My experiences with one such effort I detail in Part II. Contemplating a message that could well outlast humanity itself is sobering, frustrating, and exalting.

On shorter scales there remain enormous difficulties. Our complexity as a thinking species arises in part from the inherent conflict between the contradictory demands of these time scales. We are geared to think on the scale of $x=1$, a decade. Beyond that lies a full century, $x=2$, the boundary of posterity.

Only science and religion among the great human enterprises can endure through centuries and link us with our descendants. In the future our crucial option will be whether we use our resources to continue our present, historically extraordinary two percent growth rate per year.

There is no inherent physical reason not to expect that we will. Ambition is eternal. But to do so will enmesh us in severe crises of overpopulation and resource depletion. What sorts of "messages" can we transmit to our distant descendants in the language of the planet itself—in biological and environmental information? These themes I explore in Parts III and IV.

The future comes in all time scales, yet the cares of the day always win out over those of eternity. For example, in our unique age, growth dominates. Our population, economic resources, and sheer space packed with humans are all increasing by about two percent a year. Such population growth must end within a century, plausibly topping out at around ten billion souls.

Rather than struggles for land or riches, as in antiquity, one could agree with Dyson that "the most serious conflicts of the next thousand years will probably be biological battles." The human heritage itself could become the crucial issue. Yet in a sense this is an optimistic view, for the next few centuries promise to strain the entire human prospect in unprecedented ways. Land and riches may still be the major driver in human affairs.

Here knowledge of and intuitions about deep time can be of help, perhaps crucially. Our modern sense of time's shadowy immensities should inform our own sense of our problems. Parts III and IV consider ideas for using new kinds of deep time messages to deal with our coming crises, or at least to offset them.

These proposals shall surely be controversial. Their perspective steps upon many toes, economic, ideological, and philosophical. Deliberately writing in a tone of advocacy, I offer these ideas in an attempt to see not only far antiquity in perspective, but our times as well. Past methods of communicating across the ages had foibles and fatal delusions; ours do, too, as Parts I and II describe.

We should learn from these. Knowledge of history's panorama can aid our judgments today. Change accelerates all around us. We dwell in a unique epoch, hurtling downstream, borne by currents we can only weakly control. Only by sensing our place in the flow of time can we navigate the rapids ahead.

If we are not constrained to Earth's surface beyond the next century, our two percent growth rate per year would yield in a millennium a half-billion-fold increase in all these numbers. A message from the far past would be swallowed by such profusion, unless very carefully aimed at an audience that could not miss it. We have the examples of the pyramids and Stonehenge for strategies to achieve this: at a minimum, be big, solid, heavy, and hard to remove.

While our age offers harder materials and new locales, even the sanctuary of deep space may not remain distant in centuries to come. The uncountable numbers of lost messages should warn us that while our yearning for eternity will presumably persist, the attempt is not easy, and never certain.

How to penetrate such formidable barriers? Most of this book deals with this problem, seen from quite different perspectives, but informed by our instructive past. Peering intently toward our future, we must still cock an eye toward the rearview mirror.

PART ONE

TEN THOUSAND
YEARS OF SOLITUDE

He would lift up man's heart for his own benefit because in that
way he can say No to death.

—WILLIAM FAULKNER

One of the chores of physics professors everywhere lies in fielding telephone calls that come into one's department. Sometimes they are from obvious cranks, the sort who earnestly implore you to look over their new theory of the cosmos or their device for harnessing the Earth's magnetism as a cure to the world's energy needs. These one must accord a firm diplomacy. Any polite pivot that gets one off the line is quite all right. One of the few governing rules is that one may not deflect the call to another professor.

In 1989, I got a call that at first seemed normal, from a fellow who said he was from Sandia Laboratories in Albuquerque, New Mexico. Then I sniffed a definite, classic odor of ripe crank.

"Did I hear correctly?" I asked. "The House of Representatives has handed down a requirement on the Department of Energy. They want a panel of experts to consider a nuclear waste repository, and then numerically assess, with probabilities, the risks that somebody might accidentally intrude on it for . . ."

"That's right, for ten thousand years."

I paused. He *sounded* solid, without the edgy fervor of the garden variety crank. Still . . .

"That's impossible, of course."

"Sure," he said. "I know that. But this is *Congress*."

We both laughed and I knew he was okay.

So it came to be that a few months later I descended in a wire-cage elevator, clad in hard hat with head lamp, goggles, and carrying on my belt an emergency oxygen pack. I had a numbered brass tag on my wrist, too—"For identification," the safety officer had said.

"Why?" I asked.

She looked uncomfortable. "Uh, in case you, uh . . ."

"In case my body can't be identified?"

"Well, we don't expect anything, of course, but—" She blinked rapidly. "—you know, rules."

We rattled downward for long minutes as I pondered the highest risk here: a flash fire that would overwhelm the air conduits, smothering everyone working in the kilometer-long arms of the Waste Isolation Pilot Plant (WIPP), buried deep within the salt flat outside Carlsbad, New Mexico.

We clattered to a stop 2,150 feet down in the salt flat. The door slid aside and our party of congressionally authorized experts on the next ten thousand years filed out into a bright, broad corridor a full thirty-three feet wide and thirteen feet high. It stretched on like a dirty gray demonstration of the laws of perspective, with smaller hallways branching off at regular intervals.

Huge-bladed machines had carved these rectangular certainties, leaving dirty gray walls that felt cool and hard (and tasted salty; I couldn't resist). Floodlights brought everything into sharp detail, like a 1950s science fiction movie—engineers in blue jumpsuits whining past in golf carts, helmeted workers with forklifts and clipboards, a neat, professional air.

We climbed into golf carts with WIPP DOE stenciled on them, for Waste Isolation Pilot Project, Department of Energy. We sped among the long corridors and roomy alcoves. Someone had quietly inquired into possible claustrophobic tendencies among our party, but there seemed little risk. The place resembles a sort of subterranean, Borgesian, infinite parking garage. It had taken fifteen years to plan and dig, at the mere cost of $1.8 billion. Only the government, I mused idly, could afford such parking fees.

A HERITAGE OF WASTE

Ever since 1945, radioactive refuse has been an ever-growing problem. It comes in several kinds—highly radioactive fuel rods from reactors, shavings from nuclear warhead manufacture, and a vast mass of lesser, lightly radioactive debris such as contaminated clothes, plastic liners, pyrex tubes, rags, beakers, drills, pipes, boxes, and casings. Much of this is sitting in steel drums, many already leaking into the ground. We have run out of time.

Fifty years into the Nuclear Age, no country has actually begun disposing of its waste in permanent geologic sites. Many methods have been proposed. The most plausible is placing waste in inert areas, such as salt flats.

Also promising would be dropping waste to the deep sea bed and letting subduction—the sucking in of the earth's mantle material to lower depths—take it down. Subduction zones have a thick silt the consistency of peanut butter, so that a pointed canister packed with radioactives would slowly work its way down. Even canister leaks seem to prefer to ooze downward, not percolate back up. (One imagines a few million years later, when fossil wrist watches and lab gear begin appearing in fresh mountain ranges.)

The highest-tech solution would be launching it into the sun. Given that even the best rockets fail at least one percent of the time, this isn't a popular idea. Also, it would be far less demanding energetically to send waste packages all the way out of our solar system, as dubious interstellar ambassadors.

All these methods have good features and bad. The more active solutions seem politically impossible. Law of the Sea treaties, opposition to launching anything radioactive, and a general, pervasive Not in My Backyardism are potent forces.

The only method to survive political scrutiny is the Pilot Project, sitting in steel buildings amid utter desert waste a forty-five-minute drive from Carlsbad. The Department of Energy regards it as an experimental facility, and has fought endless rounds with environmentalists within and without New Mexico. Should they be allowed to fill this site with 800,000 barrels of low-grade nuclear waste—rags, rubber gloves, wiring, and the like? The refuse is to be packed into ordinary 55-gallon soft-steel drums, which will then be stacked to the ceilings of the wide alcoves that sprout off from the ample halls.

We climbed out of our carts and inspected the chunks of dirty salt carved from the walls by the giant boring machines. Everything looks imposingly solid, especially when one remembers that 2,150 feet of rock hang overhead.

But a central facet of the Pilot Project is that the walls are not firm at all. This Euclidean regularity was designed to flow, ooze, collapse.

We trooped into a circular room with a central shaft of carved salt. Meters placed around the area precisely recorded the temperature as electrical heaters pumped out steady warmth. The air was close, uncomfortable. I blinked, feeling woozy. Were the walls straight? No—they bulged inward. There was nothing wrong with my eyes.

Salt creeps. Warm up rock salt and it steadily fills in any vacancy, free of cracks or seams. This room had begun to close in on the heaters in a mere year. Within fifteen years of heating by radioactive waste left here, the spacious alcoves would wrap a final hard embrace around the steel drums. The steel would pop, disgorging the waste. None would leak out because the dense salt makes perfect seals—as attested by the lack of groundwater penetration anywhere in the immense salt flat, nearly a hundred miles on a side.

"Pilot" is a bureaucrat's way of saying two things at once: "This is but the first," plus "We believe it will work, but . . ." Agencies despise uncertainties, but science is based on doing experiments that can fail only in a limited sense.

Often, scientific "failure" teaches you more than success. When Michaelson and Morley searched for signs of the Earth's velocity through the hypothetical ether filling all space, they came up empty-handed. But this result pointed toward Einstein's Special Theory of Relativity, which assumed that such an ether did not exist, and that light had the same velocity no matter how fast one moved, or in what direction.

An experiment that gives you a clear answer is not a failure; it can surprise you, though. Failure comes only when an experiment answers no question—usually because it's been done with ignorance or sloppiness, so the question it asks is muddled. The trick in science is to know what question your experiment is truly asking.

Bureaucrats aren't scientists; they fear failure, by which they mean unpredictability. Here the Department of Energy threaded through a far more vexing territory: advanced technology. The Pilot Project has been held up because equipment did not work quite right, because there are always uncertainties in geological data, and, of course, because environmental impact statements can embrace myriad possibilities.

Our job was by far the furthest-out environmental impact report anyone had ever summoned forth. No high technology project is a child of science alone; politics governs. The pressure on this Pilot Project arose from the fifty years of waste loitering in "temporary" storage on the grounds of nuclear power plants, weapons manufacturers, and assorted medical sites. The simmering wastes rested in "swimming pools" of water that absorbed the heat (but can leak), or in rusting drums stacked in open trenches or in warehouses built in the 1950s.

The long paralysis of all nuclear waste programs is quite probably more dangerous than any other policy, for none of our present methods was ever designed to work for even this long. Already some sites have measured some slight waste diffusion into topsoil; we are running out of time.

Of all the politically feasible sites in the USA, the Carlsbad area looked best. Its salt beds laid down in an evaporating ocean 240 million years ago testify to a stable geology, water free. The politics were favorable, too. Southern New Mexico is poor, envying Los Alamos and Albuquerque their technoprosperity. Dry, scrub desert seems an unlikely place for a future megalopolis to sprout—ignoring Los Angeles, which had a port and ocean nearby.

The Environmental Protection Agency, which developed the current safety standards, predicted that the Pilot Project would probably not cause more than one thousand deaths over the entire ten-thousand-year span. This is an upper limit; the most likely number is much smaller, about a hundred.

Yet we had already spent well over a billion dollars to prevent those roughly thousand deaths, or a million dollars per death! This, in a country where two million die annually, and the Department of Transportation spends about $100,000 to avert a traffic death on the highways. Such matters troubled many of us, but we had accepted the job, and it *was* interesting.

So we members of the Expert Judgment Panel split into four groups to separately reach an estimate of the probability that someone might accidentally intrude into the sprawling, embedded facility.

We had intense discussions about big subjects, reflecting the general rule that issues arouse intense emotion in inverse proportion to how much is known about them. Should we be doing more to protect our descendants, perhaps many thousands of years in the future, from today's hazardous materials? How do we even know what future to prepare for?

ENVISIONING TOMORROW

We usually foresee the future by reviewing the past, seeking long-term trends. Yet this can tell us little about the deep future beyond a thousand years.

A bit over two centuries ago, what is now the eastern United States was in the late English colonial period. At least in the European world, there were some resemblances to the current world—in fact, some countries have survived this long. For this period, extrapolation is useful in predicting at least the range and direction of what might happen.

Going back a thousand years takes us to the middle of the Middle Ages in Europe. Virtually no political institutions from this era survive, although the continuity of the Catholic Church suggests that religious institutions may enjoy longer lifetimes. Most history beyond a thousand years is hazy, especially on a regional scale. Prior to the Norman invasion in 1066, English history is sketchy. Beyond three thousand years lie vast unknowns; nine thousand years exceeds the span of present human history.

The probability of radical shifts in worldview and politics means that we cannot anticipate and warn future generations based on an understanding of the past, even when we anticipate the use of modern information storage capabilities.

There are three types of future hazards. The best are those we can identify and reduce or eliminate, such as DDT and other chemicals. More ominous are those we know little or nothing about, such as some additive or emission—for example, radioactivity wasn't thought to be harmful a century ago. Finally, there are hazards we know pose deep-future hazards but which we do not wish to ban—long-lived nuclear waste and toxic chemicals essential to industry.

Instead, we decide to continue producing these, and then shove them away in some dark corner, with warnings for the unwary and unaware. Ancient civilizations did this without a thought; Rome did not label its vast trash heaps, ripe with lead and disease.

Working on the panel was intriguing but frustrating. We used scenarios to help fix specific possibilities firmly in the mind, allowing us to pick assumptions and work out their implications using common sense in a direct, storytelling way. Like extrapolating from the past, scenarios reduce infinite permutations to a manageable, if broad, group of possibilities.

A few of us referred to classic science fiction works to draw out talking points. Walter Miller, Jr.'s *A Canticle for Leibowitz*, for example, artfully showed how knowledge, passing through time's strainer, can get muddled. Watching the social scientists particularly grapple with the wealth of possibility open to them, I came to realize how rare are the instincts and training of science fiction readers.

After much wrangling, we decided to make our musings concrete by writing scenarios detailed enough to consider the physical as well as the social environment. They would also be bounded within some range of assumptions, or else the game becomes like tennis with the net down; not doing this negates the usefulness of scenarios in the first place. What bounds seemed reasonable?

Our initial assumptions were:

- The repository will be closed after the proposed period of operation (twenty-five years).
- Only inadvertent intrusions were allowed; war, sabotage, terrorism, and similar activities are not addressed.
- Active control of the site will be maintained during the "loading" and for a century after closure.
- After active control, only passive measures will remain to warn potential intruders—no guards.
- The radioactive materials will decay at currently projected rates, so the threat will be small in ten thousand years.
- No fantastic (although possible within ten thousand years) events will occur, such as extraterrestrial visits, big asteroid impacts, or antigravity.

Modern geology can yield firm predictions because ten millennia is little on the time scale of major changes in arid regions like New Mexico. By contrast, myriad societal changes could affect hazards, as readers of science fiction know well.

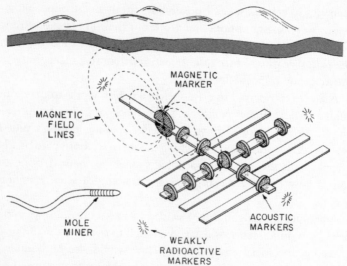

Figure 1.1 The mole miner (concept and art by Gregory Benford)

FUTURE FICTIONS

Our four-man subpanel—no women accepted the Sandia Lab invitations—worked out three basic story lines for life around the Pilot Project, based on the role of technology. There could be a steady rise in technology (Mole-Miner scenario), a rise and fall (Seesaw scenario), or altered political control of technology (the Free State of Chihuahua scenario). Envisioning these, arguing them through, was remarkably like writing a story.

THE MOLE MINER SCENARIO

If technology continues to advance, many problems disappear. As Arthur C. Clarke has remarked, "Any sufficiently advanced technology is indistinguishable from magic." A magically advanced technology is no worry, for holders of such lore scarcely need fear deep future hazards from present-day activities. Indeed, they may regard it as a valuable unnatural resource. The great pyramids, the grandest markers humanity has erected, were scavenged for their marble skins.

The societies that must concern us are advanced enough to intrude, yet not so far beyond us that the radioactive threat is trivial. Even though we assume technology improves, its progress may be slow and geographically uneven—remember that while Europe slept through its "dark ages," China discovered gunpowder and paper. It is quite possible that advanced techniques could blunder in, yet not be able to patch the leaks.

As an example, consider the evolution of mining exploration. Vertical or slant drilling is only a few centuries old. Its high present cost comes from equipment expenses and labor. An attractive alternative may arise with the development of artificial intelligences. A "smart mole" could be delivered to a desired depth through a conventional bored hole. The mole would have carefully designed expert systems for guidance and analysis, enough intelligence to assess results on its own, and motivation to labor ceaselessly in the cause of its masters—resource discovery.

The mole moves laterally through rock, perhaps fed by an external energy source (trailing cables), or an internal power plant. Speed is unnecessary here, so its tunneling rate can be quite low—perhaps a meter per day. It samples strata and moves along a self-correcting path to optimize its chances of finding the desired resource. Instead of a drill bit, it may use electron beams to chip away at the rock ahead of it. It will be able to "see" at least a short distance into solid rock with acoustic pulses, which then reflect from nearby masses and tell the mole what lies in its neighborhood. CAT-scan-like unraveling of the echoes could yield a detailed picture. Communication with its surface masters can be through seismological sensors to send messages—bursts of acoustic pulses of precise design which will tell surface listeners what the mole has found.

The details of the mole are unimportant. It represents the possibility of intrusion not from above, but from the sides or even below the Pilot Project. No surface markers will warn it off. Once intrusion occurs, isotopes could then escape along its already evacuated tunnel, out to the original bore hole and into groundwater.

This is the sort of technological trick science fiction so often explores, as a way of getting us to think about new approaches. I contributed most of this story, while the social scientists considered less optimistic ones.

THE SEESAW SCENARIO

Many events could bring about a devastating and long-lasting world recession: famine, disease, population explosion, nuclear war, hoarding of remaining fossil fuels, global warming, ozone depletion. Then the rigors of institutional memory and maintenance would diminish, fade, and evaporate. Warning markers—and what they signify—could crumble into unintelligible rubble.

Later, perhaps centuries later, society could rebuild in areas especially suitable to agriculture and sedentary life. A tilt in the weather has brought moisture to what used to be southeastern New Mexico. Farming improves, prosperity returns.

Explorers would again probe the earth's crust for things they need. The political instabilities in the region during the dimly remembered Late Oil Age had kept some of the oil from being pumped out. A quest for better power sources for the irrigation systems of this reborn civilization then leads to the rediscovery of petroleum as an energy source.

A search of old texts shows that much oil drilling had been done in the Texas territory. Since all the oil was known to have been removed from that region, explorers turn westward to New Mexico. In the spring of A.D. 5623 an oil exploration team comes upon the remains of an imposing artifact in southeastern New Mexico.

"Perhaps they left it here to tell us that there's oil down below."

"Maybe there is danger. We should consult the scholars to see if they know anything about this."

"Ah, you know these old artifacts—all rusted junk. Forget them! Let's drill and see if there's oil. . . ."

This strongly recalls *A Canticle for Leibowitz*—our "Expert Judgment" recreating classic science fiction in clunkier prose.

THE FREE STATE OF CHIHUAHUA

The year is 2583, just after a century of political upheaval in the former American Southwest. After endless wrangling caused by regional interests and perceived inequities in political representation, the United States has fragmented into a cluster of smaller nation states. Similar processes have affected the stability of Mexico, traditionally plagued by tensions between the relatively affluent North and the centralized political control of the South. Its northern provinces have formed the Free State of Chihuahua.

Political uncertainty in the Free State leads to a large-scale exodus of Anglo-Saxons, as well as many long-established Hispanic families, from the former U.S. territories. They are escorted by forces loyal to one or the other of the new countries, who practice a scorched earth policy, destroying most of the technological infrastructure, especially installations of potential military value, on the northern side of the former U.S.-Mexico border.

The Free State lacks foreign exchange and has a poor credit rating. Because it is limited in available natural resources, its people evolve into a scavenger society, recovering, repairing, and reusing all available technical artifacts from earlier times.

While excavating at the former site of Sandia Laboratory, Free State "resource archeologists" (fancily-named scavengers) discover references to the ancient Pilot Project site, including photographs of waste barrels filled with abandoned tools, cables, and clothing. They find fragmentary maps locating the site, but no references to radioactivity. In any case, social knowledge of radiation is limited, due to the development of nonnuclear energy sources during the twenty-first century—the Age of Ecology now long past.

Arriving at the site, Free State resource archeologists find the remains of markers that locate the site but do not transmit unambiguously the message that there is danger. They decide to enter. Later, the site is intentionally mined by people unaware of the potential hazard. They breach the site. Groundwater gushes up the drill, driven by the long-sealed heat of radioactive decay. Nobody can stop the gusher. A radioactive creek winds down to the riverbed, miles away.

This scenario reminds us that no nation has survived for more than a few centuries. Large states tend to fragment into smaller, more culturally coherent ones. For example, the Austro-Hungarian Empire is today divided amongst at least nine smaller countries, and something similar seems to be under way in the former Soviet Union only seven decades after its inception. Union with northern Mexico is not critical to the scenario—one can visualize a variety of ways for political control to change. As control alters, chances for inadvertent intrusion rise.

Cultural shifts can have the same effect. Suppose that in 2100 a feminist mining company comes upon our warning markers and finds them to be "another example of inadequate, inferior, and muddled masculine thinking"—and ignores them.

Other scenarios developed by the teams were quite imaginative. Suppose a Houston–to–Los Angeles tunnel breached the site? Or that a religious cult hostile to science disbelieved the warnings.

Treasure hunters might have the same skepticism. Illiterates also could fail to fathom the redundant markers. A huge spaceship in the year 11,911 could crash into the deep site.

These sketched out the vast possibilities, though the panel members assigned most of them very low probability.

Gabriel Garcia Marquez's *One Hundred Years of Solitude* alerted many of us to the subtle cultural differences between North and South America. Trying to store waste for ten thousand years of solitude reminds us, in turn, that cultural and geographical boundaries make no difference over such eras.

For example, an unspoken constraint on the U.S. program is that the waste must be stored within the country. Why not find better spots elsewhere? Mexico has many salt flats larger than the Carlsbad site. Some might well prove better, too.

The temporary constraints of politics prevented us from thinking along those lines. Under what other blinders, even unconscious ones, did we labor?

RISKS AND RELATIVES

One of the ethical philosophers on the sixteen-man Expert Judgment Panel found the prospect of international traffic in waste abhorrent. "Risk," he pronounced, "is not morally transferable."

But of course it is; we do it every day. Anyone who works in a coal mine or commutes on a heavily traveled highway incurs extra risk for some gain. Those who burn the coal or use the goods moved on the road benefit from others' activity, yet do not share the risk.

How much risk to accept is a personal decision. The ethical pivot, we felt, was that people should know the dangers they undertake.

But the Pilot Project points to a deeper problem. Over ten thousand years, no continuity of kinship or culture respects borders. Mexicans are the same as, say, New Yorkers—populations shift, societies alter. Risks resolutely kept in New Mexico are the same as risks piled up in Mexico City, for the people diffuse over these passing perimeters within a few centuries. The idea of nationality fades. Only two centuries ago Spain ruled both sites. We really are all in this together, in the long run.

This ethical relativity further questions the basic motivation in the congressional orders. Spending $1.8 billion on the Pilot Project to avert far future deaths calls into question whether we could do something more effective with the money. As panel physicist, Bernard Cohen, discussed, this brings in economic discounting of such investments. A single dollar in a trust fund set up now at three percent interest would yield in a thousand years $6 trillion!

Spending all but a single buck of the Pilot Project's price on *present* safety measures could save millions of lives over a millennium, or an average of thousands per year. With the extra dollar, we could send down the timeline bales of money to help our distant descendants. Similar investment in biomedical research now would have equally spectacular results.

Indeed, the World Health Organization estimates that fifty dollars could prevent a measles death right now in Gambia and Cameroon, or $210 per life for immunizations in Indonesia. "Since we are not spending this money to save present lives, it does not seem reasonable to spend more money to save far future lives," Cohen remarked. After all, the neighbors of the Pilot Project a millennium from now probably will be as distant from us by many measures— cultural, genetic—as the Gambians are now.

The general objection to this way of thinking is that future beneficiaries from a trust fund are not those who may be sickened by radioactive leakage. Quite so; and further, we have no way to untangle these populations.

Such arguments strongly suggest that our gut feelings do not match with classical economic discounting methods. Economists assume that an investment can carry forward undisturbed, gaining an immense multiplier effect. But never has this happened over a thousand years; banks go bankrupt, commodities crash, empires evaporate. People intuitively trust their sense of human connection more than interest rates. This is why they ignore discounting in the perspective of deep time.

This raises the question of whether there is an ethical horizon. Is there some future time, past which fretting is pointless? Physicist Hal Lewis has remarked, "No law of nature asserts that the future is less relevant than the present, but people have behaved for many centuries as if it were." Survival, he notes, is the single most important duty we have to our descendants, for without it, they won't exist. "It is pretentious to suppose that they will share our sense of values, or that we can predict their needs."

Lewis is just one of many physicists who scoffed at our congressional mandate of insuring for ten thousand years. Congress was simply adopting the Environmental Protection Agency's original 1985 regulations, but this missed significant facts.

Within a few centuries most of the Pilot Project's waste will have decayed to a few percent of its present level, though some isotopes will persist much longer. At ten thousand years it will have returned to the level of radioactivity the original ore had before it was mined for uranium; apparently, this is how Congress picked the number 10,000. Lewis points out that a single penny invested now would yield a trust fund to cover medical expenses of $10 *billion* in just 700 years. Indeed, the only way to set aside *less* than $10 billion would be to put off investing the penny until more than nine thousand years had passed. Lewis remarks, "Imagine, ten thousand years, for a country that is barely two hundred years old."

I am sympathetic to such views. Anxiety over risks from waste does not justify ignoring the simple numbers that show that we are wasting money as well.

Still, Congress wanted its estimate. We forged ahead.

MANY FUTURES

Of course, our scenarios did not exhaust all possibilities; they only sketched out the conceptual ground. Our panel also considered a "USA Forever" yarn that assumed government could indeed keep continuous control. It yielded a smaller risk, but we thought it had much smaller probability of coming true.

Such stories are fine, but how could we use them to predict quantitative probabilities? Congress wanted a number, not a short-story anthology.

We believed two elements of these scenarios most directly affect the likelihood of inadvertent intrusion: political control of the site region, and the pattern of future technological development. How could we use this intuition?

Here we used a "probability tree," which links chains of events numerically. We began by assigning simple estimates of the likelihood of single, important events. Then we multiplied these to get the total probability that a sequence of events will happen. After much wrangling, we settled on a ballpark estimate of less than ten percent chance the site would suffer intrusion.

The major risk came from the seesaw scenario of technological decline and rebuilding. Information lost because of disrupted cultural continuity meant that people could again and again make the same, simple mistakes. They might drill or mine with modest technology, unable to sense the radioactivity before they reached it, then leaving an open shaft for it to leak through.

A quick estimate can give the probability of such drilling intrusion. Research showed that the neighborhood suffered roughly one drilling per year over the last century. Assuming random drilling around the approximately four hundred square miles, the buried waste's area of about half a square mile should then have a probability of 0.001 per year of drilled intrusion. If over ten thousand years such eras occur a hundredth of the time—that is, a century in all—then there is a one percent total probability. Adding in other scenarios gives a final sum of a few percent.

Did we believe this? Of course not, in its details. There was no reason at all to prefer our guess that wildcatters would return one percent of the time. But it seemed a plausible estimate, given admittedly limited experience in the modern age.

We wrote up our result and found that the other three teams of four each had gotten the same few percent result. This added a heartening note of certainty to our otherwise rather surrealistic task. I reassured the head of the program that we could even guarantee the answer. "If there's an intrusion, I'll pay back ten times my consulting fee . . . ten thousand years from now."

Then I learned that since we finished our report first, the other teams knew our answer before they finished theirs—bad technique. A convergence of opinion is common in all prognosticating, and "experts" like us were not immune to it.

I had further worries. Physics has dominated our century, but biology may well rule the next. The implications of the Human Genome Project and rapid progress in biotechnology remind us of a more general truth: the most difficult realization about the future is that it can be qualitatively different than the present and past. This implies that an irreducible unknown in all our estimates arises from our very worldview itself, which is inevitably ethnocentric and timebound.

Are we being too arrogant when we assume we can accurately anticipate far future hazards or protection mechanisms? Probably— but we have no choice. Waste of all sorts stacks up and we must do our best to offset its long-term effects.

The Department of Energy was happy with our estimate. They and Congress could tolerate risks up to about ten percent. At present, the Pilot Project staff is finishing a trial run to further study the salt creep, how it seals, and other final technical details.

Personally, I believe the Pilot Project caverns will be filled, and that's only the beginning. Storing all our accumulated nuclear waste, not just the low-radioactivity debris the Pilot Project is designed for, would take about ten more such vaults.

What would be the point, politically or practically, in dispersing the sites? The only other site for disposal, Yucca Mountain in Nevada, is under heavy technical and political pressure. All our waste for a century could go into that single salt flat near Carlsbad.

Confining the area both lowers costs, reduces total risk, and localizes damage if it occurs. It's also politically astute. The locals want the work, and the opponents in northern New Mexico have nearly run out of legal delays. They seemed to operate out of a Not in My Backyard psychology, with no alternatives.

Part of the problem with waste of all sorts is that fears have been blown so high, few really perceive the rather minute level of risk. That was why congressional fretting over ten thousand years from now seemed so bizarre to the panel, which actually knew something about real risks. An ironic joke circulated: *How many nuclear engineers does Congress think it takes to change a lightbulb? Seven: One to install the bulb and six to figure out what to do with the old bulb for the next ten thousand years.*

During our deliberations, television stations sent their cameras and environmentalists demonstrated. I asked one of the placard-carrying men where he was from. "Santa Fe," he answered. I was surprised; he lives many hundreds of miles from the site. "They might bring some of that waste through my town, though," he said.

He was right. Spills during transport are a real, if remote, possibility. I wanted to talk to him further about sentiment in Santa Fe, which leads opposition to the site, but I couldn't tolerate his company any longer. He was puffing steadily on a Marlboro.

He could well claim that smoking was his choice, his risk—and unless he spoke out, he had no control whatever over nuclear waste. True enough—but then, there is always secondhand smoke. And the waste was generated by the federal government, an obligation settled upon all of us.

Neither Congress nor the Department of Energy has pondered the long-term issue of disposal in one site yet, but I believe it is obviously coming. The waste must go somewhere.

If we halted all nuclear power and weapons production tomorrow, we would still have a vast pile of medical contamination to care for. Nobody, I believe, wants to do away with cancer diagnostics and treatments, which produce great volumes of mildly radioactive waste.

Despite opposition, I believe eventually even local politics and bureaucratic lethargy will be unable to stop interment of wastes in the salt flats of southern New Mexico, probably before the year 2000. A government which has already invested $1.8 billion does not relish walking away from it.

Whether one regards this as a good idea or not, the political fact is that we have largely run out of time to decide how to store wastes. Holding even low grade radioactive wastes in "swimming pools," as we do now, runs real risks and can't be simply continued for, say, another century. The stuff leaks into groundwater. Increasingly, the public wants all sorts of wastes, nuclear or chemical or biological, interred far, far away from them.

Most likely, the problem will persist indefinitely. Will people give up X rays and cancer treatments? We are stuck with our largely unrecognized reach into deep time. Seemingly minor acts today can amplify through millennia, leaving legacies we do not consciously intend. Deep time has become ever-present in our age.

Given this, how will we protect future generations from such deep-future hazards, warning them about the dangerous package we've sent down the timeline? How does one send a deep time warning across ten millennia? A whole new panel pondered that question, opening up still more vexing issues.

THE EVANESCENCE OF MARKERS

The Department of Energy created two separate second panels to discuss the marker problem in detail, using necessarily science-fictional logic. I did not serve on them; the panel memberships were separate, so that the Marker panels could get some independent perspective on our recommendations. Since I knew several Marker Panel members—Frank Drake, Jon Lomberg, Louis Narens—I followed the arguments from the sidelines. Like us, they found the job to be just about the most fun possible to have while working for the government.

An illuminating moment had already come after a day of intense discussion among our so-called Expert Judgment Panel, as we considered what advice to give the Marker panels. The group spanned most sciences and was unafraid of speculative thinking. This included people like Theodore Taylor, an authority on nuclear devices, and the inventor of the 1960's Project Orion idea—spaceships driven by nuclear warhead explosions. He suggested that we detour near the Pilot Project site to find the site of Project Gnome, a nuclear test.

In 1961, Project Plowshare exploded a small warhead a thousand feet down in the same salt flat that the Pilot Project wanted to use for nuclear waste storage. The idea was to heat up rock salt and use the molten mass's residual heat to drive steam through electrical generators. It failed because the blasted-out cavity soon caved in, burying the molten salt. One would think this might have occurred to an engineer before they tried it. But that was in the golden years of nuclear development, when ideas got tried for size right away, rather than spending a decade or so mounting up piles of paper studies.

We all got out of our government-gray cars and the drivers waved vaguely at the flat scrub desert, dust devils stirring among the sage. We spread out, shooing away grazing cattle. A hoot of discovery. A granite slab, tombstone-sized, bearing a rectangular copper plaque running green from oxidation. In big letters:

PROJECT GNOME

followed by GLENN SEABORG, then director of the Atomic Energy Commission, and in smaller type the generals and bureaucrats who had overseen this failed effort.

I walked around the slab and saw another plaque, its raised lettering rusted and nearly unreadable. We could barely make out some technical detail: kilotons, warhead type, purpose, amount of residual radioactivity. At the very bottom:

THIS SITE WILL REMAIN DANGEROUS
FOR 24,000 YEARS.

If we hadn't known our quarry, we would not have found it easily out on the dry plain. Drab, small, it did not announce itself. We could tell, though, that it had been moved. Apparently, cattle needing a rubbing post had in thirty years nudged the slab several meters. How far away would it be in 24,000 years?

Ever the mathematical physicist, I reasoned that this resembled the famous "drunkard's walk," diffusionlike process. Standing in the dry desert breeze, I imagined cows bumping the worn marker in a random way. Then by standard theory, the distance wandered went up as the square root of time. If it moved, say, a meter in thirty years, then it would be about thirty meters away in 24,000 years.

Not too far to be still useful; the plaque was safe from cattle. But suppose something nonrandom decided to move it—such as a human needing chunks for a riprap wall?

This experience brought to my mind a worry: Could warning markers be self-defeating? There lurked behind our assumptions a more basic decision: whether to mark hazardous sites at all. Could the most effective warning be *no* warning?

Egypt's only major unviolated burial site, King Tut's tomb, provided us with much of the Egyptian legacy. Unmarked and forgotten because its entrance was soon buried under the tailings of a grander tomb, it escaped the grave robbers, who may well have included the priests of the time.

Could a hidden or forgotten hazard protect itself from harming future generations best of all? A "soft" surface marker that erodes in a few centuries would cover the short-term possibilities, I argued, and then avoid curiosity seekers in the far future. Without any clear sign that something important lay so far below, grave robbers of whatever stripe would have to be awfully ambitious to dig down over two thousand feet, on pure speculation. As well, high technologies would still be able to sense the buried markers, after all.

But this imposes ignorance on our descendants, who may wish

to avoid the place but not know quite where it is. Also, low-tech wildcatters drilling for scarce resources in some reemergent future would have no warning, and they might be the most probable trespassers.

Still, I proposed this notion, mostly for fun. Many of our most valued finds from antiquity were hidden, by accident or otherwise, from prying descendants. I suggested that standard-issue government concrete would be useful here: it disintegrates in about a century or so, providing everyone with a big, noticeable object for a reassuring lifetime, then erasing it.

Nobody much liked the idea, as I'd guessed. The later Marker panels rejected it, citing moral concerns, by analogy with printing warnings on cigarette packages. In the end this is a choice between two ends—knowledge against safety—mingling with the later issue of how markers might attract interest.

There was another issue, never expressed explicitly. I'm sure we had all imagined standing on a hill a few decades later, dwarfed by a grand monument, knowing that we had some small hand in putting it there. One of the major psychic payoffs in considering markers at all is the Pharaoh effect: the impulse to build a big monument to . . . well, yourself. Or at least your era. They won't forget *us* right away! Even better if somebody else (the Pharaoh's subjects, say, or the poor taxpayer) foots the bill.

MARKER STRATEGIES

Our panel, charged with estimating the chances of inadvertent intrusion into the Pilot Project salt flat buried 2,150 feet down, also suggested possible strategies for placing warning markers.

We had envisioned "miner moles" that would slowly tunnel through deep strata, searching for neglected lodes of valuable minerals. This implied a "spherical strategy"—deploying markers that were magnetically apparent from above, beside, and even below the deep repository.

The Pharaohs used one big, obvious marker for their tombs—the pyramid; we suggested as well small, dispersed tags, visible to "eyes" that could see magnetic or acoustic or weakly radioactive signs. Acoustically obvious markers could be made—solid rock unlikely to shatter and lose shape in the salt beds.

Large granite disks or spheres might be easily perceived by acoustic probes. They could be arrayed in two straight lines in the repository hallways, intersecting perpendicularly at the center: X marks the spot. These could be magnetized iron deposits, flagrantly artificial. Specially made high-field permanent magnets could produce a clearly artificial pattern, the simplest being a strong, single dipole located at the hazard's center. (This I took from *2001: A Space Odyssey* by Arthur C. Clarke.)

Radioactive markers could be left at least meters outside the bulk of the waste rooms and drifts—say, small samples of common waste isotopes. Like similar weak but telltale markers left on or near the surface, these have the advantage of showing the potential intruder exactly what he is about to get into. No language problem.

All these markers should be detectable from differing distances from the waste itself. Acoustic prospecting in the neighborhood could pick up the granite arrays. Magnetic detectors, perhaps even a pocket compass, could sense the deep iron markers from the surface. Ultrasensitive particle detectors might detect the waste itself, or small tags with samples of the waste buried a safe distance belowground. These would be small amounts, of no health risk to the curious—weaker than a radium watch, yet slowly decaying.

MONEY MATTERS

Considering vast stretches of time tends to bring on lofty sentiments. But the present is mostly ruled by money, so as an example, the Marker panels worked out the costs of erecting a Cheops pyramid, which has lasted 4,600 years. Using square blocks of granite, nine feet on each side, one could engrave all six sides with warning messages.

That way, if the exterior faces wear away, lifting one block would uncover a fresh inscription. The pyramid core could hold, not a Pharaoh, but a set of more detailed messages, for those in the future who will dig out of simple curiosity (archeologists), or those suspecting that there's a treasure in here somewhere—else why go to all the trouble?

Making all the blocks of the same material eliminates problems arising from different thermal expansion rates, which can cause cracks. Tapering the pyramid less steeply than the natural slope of a sand pile would avoid much damage from earthquakes. Like the Cheops pyramid, the load-bearing stress would be wholly compressive, using only gravity to hold it all together, with no tensile forces that open cracks.

This is expensive. If a single inscribed block costs $5,000, they would cost $62 million, about six percent of the to-date cost of the Pilot Project (though less than one percent of the projected cost over the site's entire active use).

This is no accident. Considering many different markers taught a tough lesson: longevity trades off against cost. There is no simple, good, cheap marker.

Thinking like a cost-conscious Pharaoh, suppose we make the blocks smaller, to ease assembly costs. That makes them easily climbed, increasing vandalism. It also means ordinary sized people can reach all the inscriptions without a ladder.

That opens a larger question: the biggest threat to the Pharaoh's pyramids and to a nuclear marker pyramid is pesky, grasping humans. In historic sites, metals quickly vanished and buildings were quarried. Useful, cubic blocks especially might be carted away. The Cheops lost all its cladding marble skin quite quickly; ancient Greek travelers remarked on how they could be seen as bright white beacons far across the desert, but no modern observer has found any of that left. (The Washington Monument was vandalized immediately after it opened in 1888, and the interior stairwell had to be permanently closed. Vandals don't respect greatness.)

One could offset such problems. For example, using interlocking but irregularly shaped blocks would stop their use elsewhere. Making the materials outright obnoxious might help, too—but stones that exude a bad smell steadily evaporate away, destroying the structure.

A better path might be to make the marker hard to take apart. Here the clear winner is reinforced concrete. The Cheops would take much less work to tear down than it was to build up, but the reverse is true, for example, of the Maginot and Siegfried lines of the world wars. Despite intense political pressure from local communities, the bunkers have proved to be too costly to take away. Contrast the colosseum in Rome, which has suffered greatly, with most of its building stones "recycled" into houses.

Probably the ancients understood this principle quite well, since Stonehenge (1500 B.C.) used blocks of up to fifty-four tons, and

English tombs (2000 to 3000 B.C.) used stones of up to a hundred tons. They thought the trouble was worth long-term insurance. Our experience with concrete goes back two thousand years; six of the eight Roman bridges built across the Tiber are still in service! We must be a bit cautious here, though, because it is quite possible that Roman concrete was *better* than ours.

This is because strong concrete demands a low ratio of cement to water, a very stiff mix that is tough and pricey to work with. The Romans used slave labor to ram firm concrete into place, and today's contractors pump a sloppy, muddy mix through pipes. This can make the concrete twenty times less durable than the dry, high-grade sort.

But even such precautions run into a sad lesson of history. Pyramids and other grand structures often mark honored events or people. This might be the primary message a pyramid sends: here's something or somebody important. Why not come see? And surely such a big monument won't miss this little chunk I can pry off . . .

FRIGHTENING THE FUTURE

This realization led both panels of experts toward marker *systems*— different-sized components, relating to each other so that the whole exceeds the sum of the parts. Redundancy was key. Vandalism doesn't usually take everything, so the message gets through in a holographic sense.

We have a ready example of this from antiquity. About a third of the Stonehenge stones are missing, yet we can infer the entire design without much dispute. People differ over whether there is evidence of Mycenaean Greek influence in the architectural niceties, or just what the structure was truly for, but its layout is clear.

The best way to ensure survival of truly enormous structures against both weather and pilfering is to make them out of plain old dirt; nobody will steal it. Prehistoric mounds last well. The Romans built a long wall to keep out the Teutonic tribes, stretching from the Danube to the Rhine. Even in that wet climate, while the wood is completely gone, the earth berms survive. Hadrian's wall in England is a similar case. The record is held by a chambered passage grave which is today a simple mounded earthwork in Ireland, older than five thousand years.

Both panels thought along truly gargantuan lines. Even in the far future, the old equation *big* = *important* will probably hold.

A simple berm of, say, 35 meters wide and 15 meters high, completely ringing the Pilot Project area, would demand moving about 12 million cubic meters of earth. The Panama Canal moved 72.6 million cubic meters, though of course it was an inverse monument: a hole. The Great Pyramid has the largest monumental volume, 2.4 million cubic meters. So this berm would be one of the grandest public works in history.

A far future tourist might mistake even a huge berm for a natural hill, ten thousand years from now. To get his attention, the panels wanted a ring of monoliths, probably of granite, bearing a variety of symbolic, pictographic, and linguistic inscriptions.

Stonehenge and other sites have taught us that to keep monoliths upright, more must be buried than is exposed, or else it should be firmly stuck in a rock layer below. They will probably have to be erect, too, because slanted monoliths have a poor track record. They develop tensile stresses at the surface. In brittle material like granite, once a crack develops, it reaches a critical length—and then the whole monolith splits.

There are good reasons to make none of these from composite materials—thermal stresses, as in the pyramid. This means the monoliths will be imposing, homogeneous rock, arranged in patterns that convey our message of threat.

But prudence suggests that we should also scatter small markers around the site, perhaps slightly buried, so that they get unearthed by scouring winds throughout several millennia. These would attract attention even if the monoliths somehow fail. The panels considered electrically active markers, reasoning that thermoelectric power (which would use the temperature difference between the surface and one hundred feet below) or solar power is available.

The trouble is, even the most reliable electronic components, such as those used in spacecraft or undersea cables, only last a few centuries. More reliably, we could embed contrasting dielectric materials in the site surface, which reflect radar differently. These would give a good, artificial signal to airplanes or even orbiting satellites.

We could also bury time capsules, just a bit below the usual souvenir hunter's digging zone, made of tough stuff—baked clay, tektitelike glass. These might be tablets, far better than the mud tablets the Babylonians left, quite inadvertently.

THE PHARAOH'S TOMB

Finally, everyone agreed that there should be some sort of central chamber, mostly underground, for detailed messages. Recall that the Cheops pyramid had a large, deep vault, so thoroughly looted that no record exists of what might have been there; only a large sarcophagus remains, apparently too heavy to carry away. Whatever message the Pharaoh's high priests intended to be carried forward, it says nothing to us today. Not a promising precedent, but we could think of no better strategy than deep burial beneath a striking site.

The central chamber would have a lot of plane surfaces for messages (walls and standing monoliths) and would be completely buried. It could also include buried magnets nearby, detectable with a good pocket compass even if all surface signs of the site vanish. Their fields could point at the buried waste, making the site a local magnetic locus.

But well-buried rooms would withhold their message from all but ardent study. Why not put the full description on the surface, some wondered, though perhaps in a shallow pit to reduce wind weathering?

Were we to elect to put this central room aboveground, making it readily readable in detail, there would be several durable possibilities. We could deeply chisel into granite; this was the strategy used to inscribe the Kilroy Was Here biography of a Persian king, Darius I, which has lasted over 2,500 years in open, dry weather. This boasts huge bas-relief figures carved two hundred feet above an ancient road, the first great highway advertisement, still readable. Granite is so hard to mark, vandals apparently never seriously tried.

I found it some comfort to reflect that most vandals do not like hard work, and in any case, the preferred tool of our time is the spray can, which coats, and so reduces the erosion rate, then fades away. Darius I's story had to contend with blown sand and carbonic acid in rain, but not with the sulfuric acid belched out by coal-fired plants, as now exist within a few hundred miles of the Pilot Project's site.

Stone holds messages well, since even heavily eroded lines emerge by rubbing over them on paper with a soft pencil. This worked on a faint carving of a square-hilted dagger on the inner surface of a sarsen stone, which survived the erosions of an open field at Stonehenge for perhaps four thousand years.

The oldest known and dated astronomically aligned site in the world is not Stonehenge, but Nabta, in the Sahara of Egypt. This stone circle is at least a thousand years older than Stonehenge, built about six thousand years ago. Still, the ancient passage grave at Newgrange, County Meath, Ireland, shows how some designs can survive well in fairly wet surroundings. Newgrange was made by Stone Age farmers about 3150 B.C., and though large (85 meters in diameter), it kept out water with cleverly placed grooves pecked into the boulders. This carried rainwater away from the apex, a corbel vault that did not fail, keeping the chamber dry for five thousand years.

The Newgrange site has no surviving markings suggesting its purpose, though; it is simply a grave. Contrast the fifty-ton granite slabs inside the Great Pyramid, which bore the name of the Pharoah for 4,500 years. Ireland and England are hard on even granite, so expecting detailed messages to last ten thousand years in arid New Mexico seemed plausible.

The Marker panels weighed the arguments for striking visible structures, but still felt that a buried vault would be our best bet—just what the Pharaohs chose, despite their huge investment in the glaringly obvious pyramids. A combination—something big to attract attention, a buried core to carry detailed messages—won out.

The central vault would be the most interesting and complex marker in the whole site, rugged, well-hidden, and purposely designed to be the world's longest-lasting human artifact. Undisturbed, it could last over geological eras.

If the aboveground monoliths were strikingly beautiful, maybe the locals would preserve the site because it was pleasing, protecting style rather than content—thus letting the message travel longer through time. Saving the striking, obvious structure would leave the vault below undisturbed, but also unreachable if the community protected the site from disturbance. This meant that the aboveground markers could not be too attractive. A majority of both panels favored outright disturbing, even ugly structures.

The panels began to entertain possibilities. A visitor would meet first the encircling earthworks, then a ring of monoliths—say, as wide as the length of a soccer field—and finally some central marker that would tell (or suggest) the buried chamber. The idea is to draw them in, make them feel psychologically enclosed in the monolith circle, become "involved" with the stone monuments at the center, induced to read the pictographs and messages inscribed. A tension between attracting interest then putting off further exploration was inevitable. Our detailed message was intellectual, our desired impact emotional.

TRANSCENDING LANGUAGE

Intellectuals think in terms of ideas. Most people do not.

This may be the greatest blind spot inherent in convening panels of smart folks. Often, intellect tends to frighten people away. Why not play to emotion rather than mere cool caution?

Could we convey the general *emotional* message in some direct way, independent of language?—a time-steeped semiotic agenda, as mentioned in the Introduction.

Suppose we erected some aerodynamically streamlined monoliths with gaps between them. These resonate in the wind, sending forth a hollow, mournful note. Most likely, over many centuries such wailing rocks could establish a legend about the site that transcends language. The spacings and planes of designed rock needed to produce the resonant effect could plausibly persist for at least several millennia.

There's the rub—getting through to cultures and languages we cannot anticipate. The future may see our scientific age as a passing phenomenon, an idiosyncratic momentary deflection from some One True Path we would not even recognize. So how can we expect them to share our (often unspoken) assumptions, and thus read our warnings?

Generally, we can't. But there are ways of shaping a message so it has some plausible chance of sailing intact across the great ocean of deep time. We may not be able to predict the future, but we can reach it nevertheless. One could characterize nuclear waste containers at the Pilot Project as hazardous time capsules sent into the future, not knowing where they will land or what effect they will have, hoping (perhaps even assuming) that technology will solve the problems they currently represent.

The only other alternative is to forswear hazardous technologies in the first place; but we already have plenty of waste, with more accumulating from medical uses alone, so there really is no going back.

DIGITAL IMMORTALITY

Congress mandated that the Pilot Project erect markers to warn future generations, all five hundred of them, about what lurks over two thousand feet below their boot heels. But it gave no guidelines.

I asked a computer-whiz friend how he thought we should mark the site, and he had a quick answer: "Scatter CD-ROM disks around. People will pick them up, wonder what they say, read them—there you go." After I stopped laughing, he said in a puzzled, offended tone, "Hey, it'll work. Digitizing is the wave of the future."

Actually, it's the wave of the present. This encounter was echoed by some of the Marker panels' members, making me think again of our present fascination with speed and compression as the paradigms of communication. I imagined my own works, stored in some library vault for future scholars (if there are any) who care about such ephemera of the Late TwenCen. A rumpled professor drags a cardboard box out of a dusty basement and uncovers my collective works: hundreds of 3.5 inch floppy disks, ready to run on a DOS machine using Word Perfect 6.0.

Where does he go to get such a machine in 2094? Find such software? And if he carries the disks past some magnetic scanner while searching for these ancient artifacts, what happens to the carefully polished prose digitized on those magnetic grains?

Ever since the Sumerians, we have gone for the flimsy, fast, and futuristic in communications; our fascination with the digital is only the latest manifestation. To the Sumerians, giving up clay tablets for ephemeral paper—with its easily smudged marks, vulnerable to fire and water and to recycling as a toilet aid—would have seemed loony. Yet paper prevailed over clay, so that though Moses wrote the commandments on stone, we get them on paper. Paper and now computers make information cheaper to buy, store, and transmit. Acid-free paper lasts about five hundred years, but CD-ROMs' laser-readable 0's and 1's peel away from their substrate within decades.

Music is probably the deepest method of communication across cultures. It speaks to our neural wiring, exciting pulses and rhythms that fit our mental architecture. The music of hunter-gatherer drums and pipes can instill in us feelings difficult to name but impossible to miss.

Until a few centuries ago there was no method of preserving this most airy of communications. We do not know what tunes excited ancient Rome, though we have their instruments. Our modern sound recording promises new dimensions in directing durable meanings. Except for the *Voyager* disk, which sent songs, symphonies, and shouts to the stars, this is a neglected theme in most deep time schemes; perhaps, given the speed of technological change, music is a more appropriate medium for the shorter scale of time capsules.

Still, music brings up a larger question: the mutability of all transcription, whether of the written or spoken word, or of song.

Consider the Babylonian cuneiform tablet. Many thousands of these clay bricks have come down to us, dried or fired, stamped with wedge-shaped pictographs. They are truly dead media, from the stylus, to the language (Babylonian), to the very alphabet used. Only a few hundred scholars can read them. To a lesser extent, this also applies to a papyrus scroll and a Latin incunabula on medieval theology. Already, manuals for the Osborne computer have joined this company. Media and their messages fade from our world, sometimes with astonishing speed.

A desire for truly hard copy, preferably in stone, stems from its durability. Our modern digital libraries are more vulnerable than monastic scrolls were to a barbarian's torch; a power surge and all is lost.

Worse, nothing dates more quickly than computer equipment. Already historians cannot easily decipher the punch card and tape technology of 1960s computers, and the output of early machines such as Univac are unintelligible.

Still, the future of long-term storage seems to belong to electronic media. The U.S. National Archives house about 6 billion documents, 7 million pictures, 120,000 movie reels, and 200,000 recordings. The 165-acre Library of Congress, the world's largest library, houses about 120 million items and is adding about 5 million per year. But even acid-free paper is good for a few centuries, and few books are so well published today. (Indeed, the book you are reading will probably last only a few decades.)

Recordings fade, film dissolves, even museum-quality photos pale. These perishable mountains of yellowing print imply electronic media such as CD-ROMs as the inevitable future. Even the Vatican's library, half a millennium old, is going digital.

In principle, digital is forever because it is easy to renew. Making exact copies is simple and costs much less than any other medium. But so far the burgeoning industry has not made a medium that can persist physically. Magnetic tape lasts a few years, videotape and magnetic disks at most a decade, and optical disks perhaps thirty years. So far, digital lasts forever—or five years, whichever comes first.

Even if durable, digital media have an innate translation problem old-fashioned print does not. A document's meaning dissolves into a bit-stream of electronic zeroes and ones, meaningful only to the software that made it. Stored bits can represent text, a pixel dot in an image, an audio symbol, a number. There is no way to know which, or how to retrieve it, except by reading it with the proper software and hardware.

In just the last two decades we have seen the quick-step march of mainframe computers, minicomputers, networks, and soon, optical methods. Punch cards, computer tape, magnetic floppy disks, hard disks, optical storage—what can a reader a century hence make of these? Future "cybraries" will have to contend with knowledge entombed in eight-track magnetic tapes, computer tape cartridges, analog videodisks and compact disks, plus much to come. Even when translated to new media and software, material filtering through a new format is often distorted.

Imagine how the *Iliad* would read if the only existing text of the 2,400-year-old epic had been translated into every intermediate language between ancient Greek and modern English. How much of Homer's poetry and presence would survive? The multifiltered text would be recognizable, but its essence, the spark and style and flavor of Homer, would be lost. Indeed, one might mistake it for a dry, long-winded history instead of a work of literature.

All this suggests that our recent passion for the digital is probably a passing fervor. Until it firms up into a standard method, transparent to many as text is today, with equipment that promises to survive a few human lifetimes, it seems an unpromising way to consign one's vital messages to the abyss of centuries.

Eventually, neither paper and CD-ROMs, nor any foreseeable computer-based method, are for eternity. Even tombstones blur, and languages themselves are mortal. How to talk across the ages, to call out a warning? How even to get their attention? We have to learn to write largely, clearly, permanently. And largely may be most important of all, for the crowded human future may well drown out the softer voices, whispering of the distant past.

More deeply, how do we induce respect for whatever warnings we leave? Nobody will revere small, digital records, so they should be associated with larger, striking monuments. The Marker panels seemed to me to want a very special response: not the grudging respect accorded an ancient threat, but a reflective consideration.

Buildings of religious, emotional, or memorial impact tend to fare well. Cemeteries, for example, can hold their own against urban encroachment. One of the striking images as one approaches Manhattan from Laguardia Airport is the broad burial grounds, still there after centuries despite being near some of the world's most valuable real estate. In Asia and Europe, temples and churches survive better than the vast stacks of stones erected to sing the praises of more worldly power.

Of course, often they were better built, but as well, communities are hesitant about knocking them down. Often, new religions simply adopt the old sites. The Parthenon survived first as a temple to Athena, then as a Byzantine church, later a mosque, and now it stands as a hallowed monument to the grandeur of the vanished Greeks who made it.

Sometimes conquest destroys even holy places, as when the Romans in A.D. 70 erased the Temple of Solomon. Perhaps some conqueror thousands of years from now would pass by the Pilot Project monoliths, berms, and buried rooms—if, indeed, the rooms haven't been exposed, turned into a tourist attraction. Seeing them as tributes to a society now vanquished, he might order them all knocked over, buried, their messages defaced. Something comparable happened many times over as the Europeans moved across the planet a few hundred years ago, rubbing out the religious and literary past of whole peoples. The Mayans wrote on both paper and clay, but nearly all of their work is gone.

In this perspective, digital storage has a trump: make many copies, so even fanatics of the future cannot find them all. Scatter them.
Leave the translating to an ingenious future, as all antiquity did.

EYES OF THE FUTURE,
CUES OF THE PAST

Our charge from the Department of Energy was to consider inadvertent *human* intrusion into the Pilot Project. An important ad-
jective.

I personally do not think the human species will remain intact
for even the next thousand years, much less the next ten thousand.
Unless we soon halt progress in biotechnology, and don't recapture
the ability to tinker with our own genes, I expect that variants on
our Cro-Magnon theme will appear.

Other posthuman species will have ways of thinking quite different from our own. Still, even if they have extensive physical
modifications—one finger like a screwdriver? a stomach that can
digest cellulose into sugars? a better designed back?—I expect they
will share the deep programming we primates picked up far back
there on the African veldt.

Among that ancient legacy is a set of preferences for particular
landscape features. Universally, we share likings because they were
adaptive. Such "landscape archetypes" may well be so strong because Darwinnowing for them covered many hundreds of
thousands of years, as small hominid hunter-gatherer bands made
their way across rugged terrain. Developing consciousness got im-
printed in neural architecture, as the whole primate mind-body
integration proceeded with dazzling speed.

Tied every moment to weather and the wiles of other species,
our ancestors sensed themselves as part of a living unity. This legacy
had and has powerful spiritual impact. We often equate the natural
with the holy. Our enormous emotional ties to that view may seem
a form of nostalgia, but they are no less powerful for their distant
origin.

Prehumans who preferred the savannah prospered; seemingly,
those who preferred swamps or highlands did less well, since fewer
of us now prefer those places. These "hard-wired" preferences have
little survival value today; residents of Colorado do not live shorter

lives than those of Tennessee. Still, in our cerebral cortex, the past shouts and the future can only whisper.

The biologist John Appleton believes three types of cues rewarded prehumans who could pick them up: hazards, prospects, and refuges. Hazard-rich images or smells reach right into the brain, arousing anxiety that can only be resolved by taking action: the flight-or-fight response.

Taking action relaxes us, may even bring pleasure. People heavily into this go to scary movies or ride roller coasters, and get a genuine, evolution-ordained kick out of it. Most of us simply prefer landscapes we recognize, that balance prospect (views) and refuge, and invite exploration—that is, aren't boring.

MYTH

This kind of thinking goes further, into the murky realm of mythic consciousness.

Presumably our evolutionary record is written into our basic internal stories, because once these tales were true. They sit down in the unconscious, ready to spring out and make surrounding events coherent. Candidates for these are father, mother, authority, self, childhood, femininity and masculinity, gathering food, eternity, circles and squares (Plato's divine forms, somehow useful back on the savannah), devil/evil, god/goddess/good (note the similarity of these words even in as advanced a language as English), sleep, pain, death, communion. I would add number, space, and time—but then, I'm a mathematical physicist. These may be the substratum of human experience, how we construct meaning, whether it be in myth, language, religion, art—or artifacts.

Joseph Campbell became famous for popularizing the species-wide myth themes: virgin birth, the great mother, the creation of All from a chaos of nothing, the fire-theft, the plentitude of Eden and the beauty of paradise, the return of chaos in flood or deluge, the land of the dead, the dying and resurrected hero/god, the great quest journey, the sacred versus the profane, redemption through suffering and sacrifice.

We extract these stories from our environment because we are hard-wired to "see" them popping out, patterns that order a chaos. The argument here is that what seems to us to be meaning *in* the world is in fact our projection of meaning *into* the world. But all this came from the utility of such filters that sort out a savannahlike plain into easy categories.

Most of the Team A panel felt that warnings should rely on an archetype strategy, as championed by Michael Brill: ". . . archetypes are ones that, when experienced 'in the beginning,' already had a mythically significant 'tone.' " These views he reflected in some of the figures given here.

Team B mostly disagreed, championed by Jon Lomberg. This surfaced in discussions of using artful designs. Both panels could agree on physical aspects, such as the use of granite, berms, and enclosing geometries, but predictably differed over the more murky aspects of social impacts.

In the end, Sandia Laboratories rejected relying heavily on archetypes in their final report, adopting instead a direct communication strategy incorporating the use of monoliths bearing messages in many languages, plus symbols and pictographs. The essence of their method lies in the details, many of which have broad implications for all deep time markers.

MARKERS WITH MESSAGES

Following work of anthropologist David Givens (now of U.C. San Diego), the Marker panels noted that there would be four levels of knowledge to convey at the site:

Level 1: Simple ("humans made this; it's important")
Level 2: Cautionary ("trouble!")
Level 3: Basic ("this is old and technical")
Level 4: Detailed ("these radioactives are here—leave alone")

The first is essential, because the others emerge only if the site clearly appears artificial. Seen from eye level, the whole pattern should strike one in a single glance: an attention grabber. (The huge stone circle at Avebury, not far from Stonehenge, is not widely known because its stones are small compared to the circle, so one can stand in it and not realize the whole design.)

Each level should call forth a more sophisticated response. We had to count on the viewer realizing at some point that expert knowledge might be needed to comprehend the site. The four levels had to feed into each other, calling attention to themselves, holding attention and leading the visitor on.

Further, the site will compete with a plethora of all present monuments and an undoubted plentitude to come: statuary from the Civil War and wars to come, stumps from abandoned freeways, the carcasses of banks and stadiums, the tomb of Marcussa the Great, who ruled Higher Novo Mexico from 2348 until her untimely death in 2472 . . .

We also wished to convey an uncommon message. Most monuments proudly announce that Kilroy Was Here, or This Is High Church; in both cases, there is an implicit message: remember us, and so pay respect. The Pilot Project is self-effacing: we were here, so stay away.

Keeping such constraints in mind, the Marker panels dreamed and doodled sketches. Some basic designs emerged.

To honor important people or events, we erect beautiful, soaring monuments that mirror our aspirations—the pyramids, Cleopatra's Needle, the Washington Monument, even the monolith in *2001*. The waste site has to send the opposite message, straight into the collective unconscious, drawing the eye yet repelling the spirit. A model might be the Berlin Holocaust memorial zigzags, its hard edges offering no comfort or nobility.

Panel A tried to envision how the complex would appear to someone approaching in the far future.

The Plain of Thorns would sprout eighty-foot-high basalt spikes, erupting from the ground. They jut at all angles, adding to the sense of jarring discordance, though such steep angles can cause cracking and faster erosion, because the cantilevered shafts under tension develop minute cracks. To offset this problem, perhaps use a field of spikes, perfectly vertical. Or blend the ideas, with spikes interspersed among the thorns. If the Thorns can't fall and damage the spikes, eventually only the spikes remain, in a field of rubble.

Another possibility is the Black Hole: a black basalt slab, unbearably hot from accumulated solar heat. Laced with thick, crazy-quilt expansion joints like cracks in parched plains, it forbids farming or drilling. Its heat may reduce surrounding vegetation and make it easier to detect from the air at night by its infrared emission.

Or consider the Rubble Landscape: the local stone, dynamited and bulldozed into a crude square pile covering the whole project.

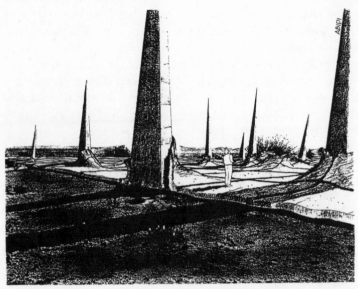

Figure 1.2 Mixed thorns and spikes (concept by Michael Brill, art by Safdar Abidi)

Figure 1.3 Spikes bursting through a grid field (concept by Michael Brill, art by Safdar Abidi)

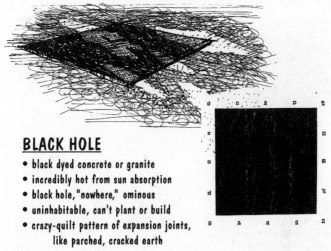

BLACK HOLE

- black dyed concrete or granite
- incredibly hot from sun absorption
- black hole, "nowhere," ominous
- uninhabitable, can't plant or build
- crazy-quilt pattern of expansion joints,
  like parched, cracked earth

Figure 1.4 The Black Hole, seen from the side and directly above (concept and art by Michael Brill)

It rears above the landscape; hard to hike through, a place destroyed, not made.

With a bit more trouble, one could make Forbidding Blocks: that same broken stone, cast into mixed concrete/stone blocks twenty-five feet on a side, dyed black, irregular, distorted. They define a square, with "streets" five feet wide between blocks. But the streets lead nowhere and no one could live or farm there. The blocks get very hot, and the whole crudely ordered array massively denies use. Some granite blocks stand out, covered with inscriptions, warnings.

The favorite of many panel members was the fifty-foot-high Menacing Earthworks, all radiating outward from the bare site center. These are lightning-shaped, jagged, crowding in on the tiny traveler, cutting off views of the horizon, chaotic. At the open center is a world map and a crosshair fixed on the site's location. At the intersection also lies the Pilot Project concrete hot cell, housing small samples of the interred waste, so even a simple Geiger counter can show that they are radioactive. The hot cell is slowly going to ruin.

The Menacing Earthworks central area is typical of the other designs. Beside the hot cell lies a vast walk-on world map of all

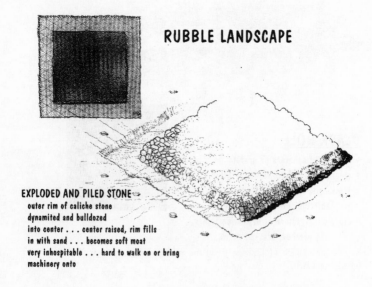

RUBBLE LANDSCAPE

EXPLODED AND PILED STONE
outer rim of caliche stone
dynamited and bulldozed
into center . . . center raised, rim fills
in with sand . . . becomes soft moat
very inhospitable . . . hard to walk on or bring
machinery onto

Figure 1.5 The Rubble Landscape (concept and art by Michael Brill)

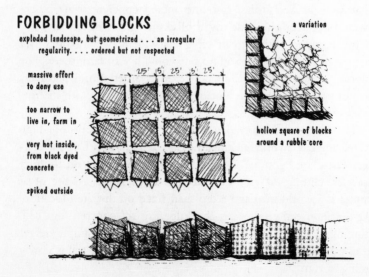

FORBIDDING BLOCKS

exploded landscape, but geometrized . . . an irregular
regularity. . . . ordered but not respected

a variation

massive effort
to deny use

too narrow to
live in, farm in

very hot inside,
from black dyed
concrete

spiked outside

hollow square of blocks
around a rubble core

Figure 1.6 Vertical and side view of Forbidding Blocks (concept and art by Michael Brill)

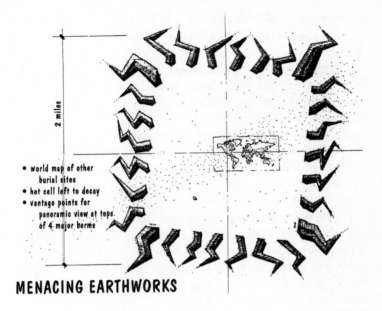

- world map of other burial sites
- hot cell left to decay
- vantage points for panoramic view at tops of 4 major berms

2 miles

MENACING EARTHWORKS

Figure 1.7 Menacing Earthworks seen from above (concept and art by Michael Brill)

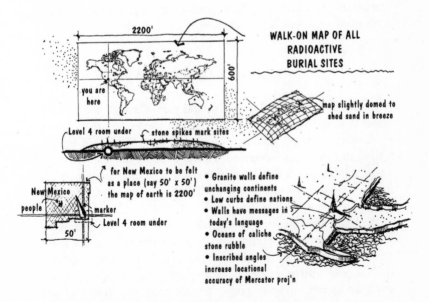

WALK-ON MAP OF ALL RADIOACTIVE BURIAL SITES

2200'

600'

you are here

Level 4 room under — stone spikes mark sites

map slightly domed to shed sand in breeze

New Mexico

people

50'

for New Mexico to be felt as a place (say 50' x 50') the map of earth is 2200'

marker

Level 4 room under

- Granite walls define unchanging continents
- Low curbs define nations
- Walls have messages in today's language
- Oceans of caliche stone rubble
- Inscribed angles increase locational accuracy of Mercator proj'n

Figure 1.8 Walk-on map of the world's radioactive burial sites (art by Michael Brill)

repositories of waste. Added is a map of New Mexico showing this site. The map is of granite and domed, so sand blows off and rainfall cleans it.

A kiosk atop the map carries the Level 2 and 3 messages. Its heavy concrete and granite should endure ten thousand years.

Beneath, a buried room holds details about what lies in the salt bed below, as do four smaller buried rooms beneath the largest earthworks. Inscribed "reading walls" of granite appear throughout the site, standing at different angles so that wind erosion cannot deface all of them. The Level 4 message includes three-dimensional drawings of the entire facility, including the corridors cut far below in the salt bed.

The common ideas here are a forbidding prospect, irregular geometries, and anticraftsmanship. This contradicts human archetypes of perfection in our imperfect world, which circles, squares, pyramids, and spires would echo. Using crooked forms when plainly the designers knew "better" suggests a deliberate shunning of the ideal, a lack of value here. *Bad ju ju.*

People value craft, too, so these designs are roughly made, of materials such as rubble and great earthen mounds that discourage workmanship. Yet they are large, important—suggesting that there is no pride or honor here, but there is sobering danger.

MESSAGE/MEDIUM

A resounding emotional theme should echo through the inscriptions: awe, apprehension, outright fear—independent of language or culture. In simple drawings one could show anguished human figures and especially faces, made clearer by using bas relief. Perhaps a repeated theme: a face with hands, sculpted in abject horror, as in Edvard Munch's well-known painting, *The Scream*. Or perhaps an eloquent warped face, nauseated.

With the wind blowing through the monoliths, coaxing mournful acoustic resonances from their curves, a dissonant and wailing aura should surround the place. Whatever cultures come and go, they should inherit a legend of a spooky, disagreeable place—whether or not anybody knows exactly why it is that way any longer.

These details would await the intruder who digs. Each marker

MESSAGE KIOSK

- For Level 2 message in 7± languages, Level 3 in several, and blank areas for reinscription in current languages
- Concrete "mother" wall protects granite message wall from wind driven sand erosion

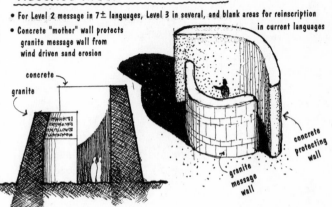

Figure 1.9 Reading walls and message kiosk (art by Michael Brill)

BURIED ROOM: for Level 4 messages

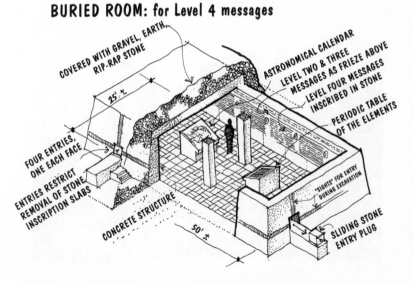

Figure 1.10 Buried Level 4 messages (art by Michael Brill)

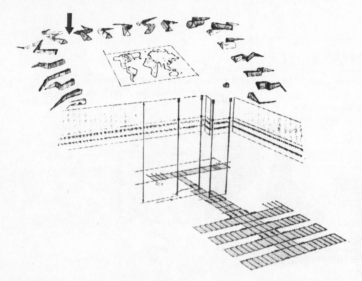

Figure 1.11 Perspective view of the site showing the waste corridors cut in the salt bed below, with elevator shafts and marker features visible. The arrow points out the reader's location at a Level 3 message center.

Figure 1.12 Confronting the Forbidding Blocks (concept by Michael Brill, art by Safdar Abidi)

design had a buried room at the center. Therein would lie plenty of duplicate technical detail, from lists of radioactive elements in the site to a periodic table of the elements itself, for correlation with the notation on the walls.

The buried vault might be plundered. Though we would take pains to make this difficult, the high priests took such measures, as well, with little success. How to insure against pillage?

Here the Sumerians left us a valuable lesson. Around the third millennium B.C. they began writing on little clay tablets, letting them pile up in large numbers to reach us in the Christian era. This left us an unbroken line of hard documents with dazzling detail about religion, beliefs, economics, customs. Similarly, we should seed the waste site with small, ceramic plates, carrying compressed warnings and information. This could offset vandals who wreck the big, imposing monuments, or natural disasters. As erosion changes, buried plates get exposed: time-released information.

If they can read them. But our current languages are not going to make it on the long march of millennia.

MUTABLE TONGUES

Languages change unpredictably. They are so complex that tendencies to simplify one part—say, in grammar, when English shed the masculine/feminine/neutral articles and verb forms—will quite likely trigger complication in another (in English, more irregular verbs).

Historical accidents bring great change. The main reason that English differs so profoundly from its closest German relative, Frisian (spoken in the northern Netherlands) is that the Angles were invaded by French-speaking Normans, and the Frisians were not.

No artificial language can avoid such evolution by chance, either. Esperanto, which once had about fifty thousand speakers, was effectively blocked from official usage when the U.S. and U.S.S.R. vetoed using it as the working language of the U.N. It has more speakers than ever today, but nobody tries to get around our world by using it. Seen a few centuries hence, our age (say, 1700 to 2100) might be termed the Anglo-Saxon Era: when the language, literature, political, and economic tenets of a tiny island spread over the entire globe, uniting it under a cultural umbrella. Against the inexorable rise of English, Esperanto had no chance, however rational its use may seem.

Languages evolve, just as in biology, by descent and divergence. We can read Shakespeare but need notes to fathom some of his archaic words. Without substantial help, *Beowulf* is beyond us. As a rule of thumb, basic vocabularies change about twenty percent in a thousand years. Even if a language survives (as most don't) for five thousand years, it will be vastly different.

Realizing that there will never be a science that predicts future languages, the problem of writing in the Pilot Project markers becomes central. After a few centuries, only experts can read earlier forms of their own language. With help we can struggle through the original Chaucer, but *Beowulf* is marginal at best. If there is no great cultural discontinuity, probably a few antiquarians will be able to decipher our English or any other current tongue. But antiquarians seldom consort with vandals or wildcatters.

Though finders of these many buried messages may fail to read the languages inscribed, they might recognize a symbol. Our evolutionary legacy gives us some predispositions to seeing gestalt wholes, so we naturally group objects enclosed by a line. We're sensitive to edges, and readily pick figures out of ground. Breaking down information from large chunks into bits comes easily to us.

Symbols should play to this. We like narratives; the easiest way to teach another is to tell him a story. Narrative history can be dated to at least 11,500 years ago, when the big explosion of human sign artifacts began with Spanish Levantine rock art. These were pictographs showing hunters, weapons, clothing, prey, with both sexes as players.

At the waste site, similar simplified line drawings could show stick figures burying the waste, warning others away. Others could present people digging or drilling into the site; groundwater flushing up through the hole; a man getting sick, falling down, dying, then others mourning him.

The story should unfold in different ways, touching on the great mythic stories where possible. Our tale may become a legend, as well, if its story line is clear and striking. The Bayeaux Tapestry of twelfth century France, the Japanese scroll of *The Mongol Invasion* and the Lakota Sioux picture story *The Battle of the Greasy Grass* (to us, Custer's Last Stand)—all gather their power through successive images.

Storytelling is a powerful current linking eras. Why not use the oral tradition of the region to carry our warning? The *Iliad* and the *Odyssey* of Homer made their way to us through millennia as purely spoken stories, after all. Even after they were written down in the sixth century B.C., the final text did not settle down for four centuries. A great saga commissioned to lend mythic status to the Pilot Project might do just as well.

But of course, nobody can reliably order up mythic works. Even if the work survived, told and retold, it would evolve, maybe lose its essential warning function. And experience shows that once oral traditions get written down, they fade as great tales. Books entomb storytellers. Homer's *Odyssey* lives in print, not in rolling spoken poetry.

PICTURES

So we are left with telling a picture story. While a picture may be worth a thousand words, there's always the problem of knowing *which* thousand words are evoked. In just this century, the swastika went from a positive religious symbol of India to the hated Nazi emblem. We want to call across the millennia, "Poison! radioactive materials—don't intrude."

The panels considered our most common symbol for radioactive materials, the "uranium atom" of three ellipses centered on a dot. But this merely describes, doesn't warn. And some people mistake it for a solar system.

Our current "radiation" symbol is international: a "trefoil" of a black circle with three vanes sticking outward. Globally used, it might survive many centuries. But it is not an icon; it is just an arbitrary design, and nothing about it intrinsically relates to the idea of radioactives. Some see it as floral design or like a Japanese *mon*, a clan crest. One team member quipped, "Ummm—why are they burying all those submarine propellers?"

The skull and crossbones go back to medieval alchemists, who saw in it Adam's skull and crossed bones promising resurrection. Only later did it come to mean poison, and though it is international, it has problems. In an experiment with three-year-olds shown the symbol, they immediately shouted "Pirates!" Put it on a bottle and they shouted "Poison!"

Still, skulls alone have a powerful, universal horrific effect. This may spring from our origins, when the sight of skulls warned that predators that preyed on primates were about. Shelley's poem "Ozymandias" invokes scattered body parts, an unsettling motif, and the disembodied head recalls the deep fear primates display near skulls. There is an easy sociobiological explanation for this: predators on early primates would tend to leave heads in their wake (difficult to eat), and automatic fear of these leftovers would prod other primates to quick evasion. Generally, though it seems ghoulish to think so, scattered primate parts might be the most lasting warning label. The human visage rendered down to bare bone summons up the specter of death, and will probably do so for a long time.

Though no symbol will probably last ten thousand years, perhaps a cluster of them would help. The "Mr. Yuk," a recently adopted poison warning, is a Happy Face reversed into a scowl, tongue sticking out, eyes squinting. Put that together, say, inside a slashed circle. To show that the slash means negation, we could display other X'ed-out symbols. But what to X out? A drilling tower?— but in profile it is easily mistaken for a monument itself. A pictograph stick figure digging? This doesn't hit the mark because in fact nobody could reach the salt bed that way; we can ignore weekend curiosity seekers. No single idea quite works.

Even if our markers do make it through to the far future, how could we convey the huge expanse of time we had mastered? Various schemes for using astronomical "clocks" could work, as in Figure 1.13, which uses the motion of the apparent north pole in the sky to measure millennia. Combined with faces and a radiation symbol, the general gist and feel for the time scale might be explicable.

The panels urged attention to erosion. Figure 1.14 shows how a berm site could turn a clear trefoil symbol to a scattered, blurred image, as seen from a nearby height. Among this eroded site stand message markers like Figure 1.15, bearing detailed data and warnings like Figure 1.16.

A more subtle problem is that exposure to radioactive materials usually takes many years to do harm. Much damage could occur before anyone made the connection to a breach at the site. One possible way to convey this is to tell a story, starting with a child figure encountering the waste (represented, say, by the trefoil).

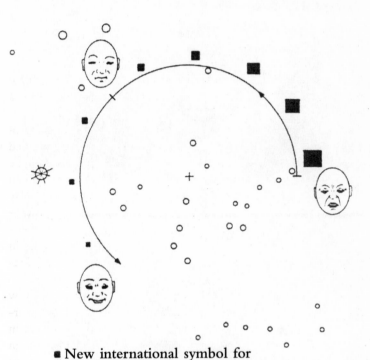

■ New international symbol for
"radioactive waste buried here,"
or standard trefoil.

Figure 1.13 A Level 3 diagram, based on the movement (precession) of the North Pole. As the Pole moves in the night sky from the right (near the star Polaris and the frowning face) the size of the black blocks (standing for an as-yet-undecided symbol for radioactivity) shrinks, suggesting the waning of danger. The face becomes neutral after the Pole has moved through ten thousand years, and happy still later. (Faces are based on figures from Iranaus Eibi-Eibesfeldt; copyright 1989 from *Human Ethology*, Aldine de Gruyter, New York.)

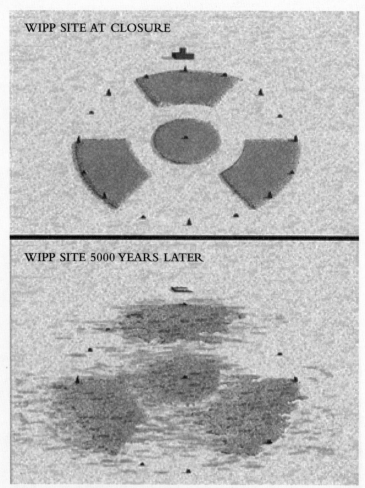

WIPP SITE AT CLOSURE

WIPP SITE 5000 YEARS LATER

Figure 1.14 A site whose berm markers form the radiation trefoil. At top, at the future sealing of the site; below, after five thousand years. (art by Jon Lomberg)

Figure 1.15 A tall granite monolith with inscriptions (art by Jon Lomberg)

WARNING! Radioactive waste buried here.

Poison! No drilling, no mining, no explosives!

אזהרה : פסולת רדיואקטיבית קבורה פה

אסור לקדוח, אסור לחפור , אסור

¡**PELIGRO!** Basura radioactiva enterrada aquí.

¡Veneno! No hoyos, no excavación, no explosivos.

Figure 1.16 Warning signs using the same symbol(s) and different languages. This Rosetta stone approach will convey that all languages carry the same meaning. Team B proposed including major modern languages, liturgical ones, and the languages of peoples native to the region. (art by Jon Lomberg)

Next comes a boxed cartoon with the symbol now on the figure's chest and young, short trees nearby. Next panel, the trees have grown and the child is an adult—lying down, scowling, feeling bad. Simple, direct—See Dick Run from Radioactive Death.

These may help convey meaning after all language connection with us is lost. A few antiquarians may know how to decipher the inscriptions, but wildcatters won't necessarily call on a distant university for help.

LASTING BEAUTY/AMBIGUOUS ART

Some Marker Panel members felt that while the monuments should be discordant, to carry the essential threatening message, they should have some aesthetic appeal. "Beauty is conserved, ugliness discarded," one said. The pyramids may have survived in part because they are striking, not just because they are bulky and hard to tear down. The same might prove a useful strategy for the Pilot Project markers. "A gift from our century to the future," one suggested. Another proposed commissioning artists for a large scale environmental sculpture.

Trouble lurks here, I feel. So did panel member Jon Lomberg, who with Carl Sagan and others designed the messages sent on board the *Voyager* spacecraft to the outer solar system. Even if *we* think our markers are ugly, he said, "How can you be sure it won't be mistaken for art?"

Art is ambiguous. As a universal language it tells little of the artist's intent. Cave paintings of animals don't tell us why they were made. Representational art fares better than symbolic, but the marker designs were quite symbolic, as is most large scale sculpture. Recall how often you've heard viewers puzzle over the intent of abstract painters.

Further, said Lomberg, "Even if we could commission some monument great enough to become a wonder of the world whose fame would be carried down through three hundred generations, the very fact that the marker was so impressive could lead to the belief that the purpose of the marker was artistic rather than communicative." A big, powerful sculpture isolated amid desert wastes could be seen as like Mount Rushmore, a spot with a sole, uplifting message.

Art often has no function; it is an experience, period. Even art trying to be ugly, as with the fearful faces, could miss its supposed target. Picasso's *Guernica* wasn't really warning us to avoid the Spanish Civil War.

Worse, art draws a crowd. "We want people to stay away from this site, not travel from distant places to see it." Suppose it is so striking that in a few millennia it draws tourists who come to see the ancient wonder. Perhaps by then it is the oldest surviving large monument, ever since a revolutionary regime in Egypt took the pyramids apart for a People's Residence. The tourists need a hotel to stay, which needs water, so it drills . . .

Some felt this was a remote worry, arguing again that beauty survives better than ugliness. But does anyone expect that our government can commission great art? It has enough trouble agreeing on mildly interesting but intensely controversial photographers and performance artists. Lomberg remarked that for every successful commission there are a hundred failures, from the Prince Albert Memorial in London ("an architectural laughingstock") to the Airman's Memorial in Toronto, locally known as "Gumby Goes to Heaven."

Lomberg pointed out that much of the art world is antiscientific, antirepresentational, and favors detached, nihilistic work. He doubted that our present art community would be well-qualified to create or even select a design that was informed about the many scientific and technical intricacies needed—aspects like encroachment of sand dunes, material durability, future technologies. Announcing a grand competition for ideas virtually promises that something will be chosen, adequate or not. "They're likely to end up picking a giant inflatable hamburger to mark the site."

Suppose further that this Pilot Project does turn out to be the model of future sites. Will the French or Chinese use a marker system—symbols, art, and all—like ours? Or will national rivalry rear its head? If so, then two thousand years from now it will be hard to see that these variously designed places scattered around the world have some common story to tell.

Thus the present, irascible humanity of us all could well propagate into the far, far future. (Indeed, perhaps this is a deep, unspoken motivation, a blend of High Church and Kilroy Was Here.) The Pilot Project caverns will not be filled and need marking until around 2030, but thinking about it has begun precisely because we need to mull our way into that inconceivable perspective—a time when we will not merely have vanished, but probably our entire culture as well. This is the first radioactive sepulcher in the world, and may set the standard for all others, nuclear or otherwise.

UNHOLY TRINITY

In closing, consider Trinity Site, where the first A-bomb exploded in 1945. At Ground Zero in White Sands, New Mexico, the blast left a glassy crater of fused aluminum silicates a quarter mile across and twenty-five feet deep.

Now there is nothing. Dry winds filled the crater; tough desert plants had cracked it. Radiation levels are only very slightly higher than the background of ordinary scrub desert. Life had reclaimed its territory in a single human generation. The "message" of Trinity is gone.

The easy problem of deep time is time's rub. Greater still is the abyss of culture we must cross. The barren Trinity site recalls Shelley's great "Ozymandias":

I met a traveler from an antique land
Who said: Two vast and trunkless legs of stone
Stand in the desert. Near them, on the sand,
Half sunk, a shattered visage lies, whose frown,
And wrinkled lip, and sneer of command,
Tell that its sculptor well those passions read
Which yet survive, stamped on these lifeless things,
The hand that mocked them and the heart that fed;
And on the pedestal these words appear:
"My name is Ozymandias, King of Kings:
Look on my works, ye Mighty, and despair!"
Nothing beside remains. Round the decay
Of that colossal wreck, boundless and bare
The lone and level sands stretch far away.

Deep time is as much the province of the poet as the scientist.

I came away from my experience on the Intrusion panel quite sobered. Standing in the shadow of such vast issues brought home my own inadequacy, faced with so many varied cultural and scientific questions. Many questions arising in the course of our work demanded a broad, skeptical learning none of us embodied particularly well. Important issues could well have fallen between our professional chairs. Still more were to emerge in my next engagement with deep time messages.

The nuclear waste markers will be our society's largest conscious attempt to communicate across the abyss of deep time. There will be others, for the problem of virulent waste will not go away. Indeed, given our rising population and greater control of resources, waste shall probably mount for the foreseeable future.

How we present ourselves in these ancient sepulchers may be our longest-lasting legacy. It is sobering to reflect that distant eras may know us mostly by our waste—and by our foresight.

PART TWO

VAULTS IN VACUUM

*Nothing is more complex than the unknown, for it masquerades as
the unknowable.*

—MARK O. MARTIN

The central images of the 1968 classic film *2001: A Space Odyssey*
revolve about a mysterious message left in the form of a monolith
buried on our moon. It had been waiting for millions of years for
us to show sufficient ability to uncover it.

Deep space is indeed the best place to leave a long-term message
that can survive in the high vacuum and isolation available beyond
Earth: deep space equals deep time. Soon after the advent of the
space program, scientists proposed sending messages aboard spacecraft.

The first concerted attempt to send a material message beyond
Earth rode upon the first spacecraft to leave our solar system, *Pioneer*
10 and 11. Launched in 1971 and 1972 to fly by several outer
planets, their support struts carry a six-by-nine-inch, gold-anodized
aluminum plaque. Each bears an etched drawing describing some
facts about our civilization. A sketch of two nude humans, greeting
the infinite with a hopeful wave, became its best known feature.

In 1977, NASA launched the *Voyager* missions to the outer planets, each bearing an Interstellar Record created by a team including
Carl Sagan, Frank Drake, and Jon Lomberg. The phonograph record (already an obsolete medium) carried both sights and sounds of
Earth, from Gregorian chants and whales to Chuck Berry, and set
the standard for broadly based, information-dense messages. Other
small messages—a microdot of inscribed names on the *Viking*

lander to Mars, and an honorary plaque on the failed Russian Phobos mission to Mars—added nothing new.

More than a decade passed before another substantial attempt. The Russian Mars '94 mission collaborated with the Planetary Society, a private advocacy group, to affix a compact disk (a CD-ROM) inside each of two landers that would descend to the deserts of the Red Planet. Intended as a gift to future explorers, a compact disk bearing text, audio, and images of Mars rode inside the lander, with an exterior label attached to a small petal outside.

The label was a diagram showing how to find the compact disk, plus instructions on how an aluminum-gallium-arsenide laser reads the disk surface and a computer decodes and shows the contents. At the center of a silicon chip on the plaque were instructions for playing the disk in letters one micron high (a thousandth of a millimeter).

Also attached was a "nanolith microdot" termed the MAPEX, for Microelectronics and Photonics Exposure experiment. This sandwich of solid components was meant to be retrieved eventually and used to measure effects of prolonged exposure to deep space on special materials. The disk was made of silica for durability, as the designers assumed at least a century would pass before any person or robot could recover it.

I became involved late in the program when asked by the Planetary Society to work with the disk's producer, Time Warner Interactive. The project director for the CD-ROM was Jon Lomberg, who had a long history with spacecraft messages and the Pilot Project Marker panel. CD-ROMs gave him enormously larger space to send digitized cultural works.

The CD-ROM disk opens with photographs and short quotations by Robert Goddard and Konstantin Tsiolkovskii, the two great pioneers of spacecraft. Writings by scientists and science fiction writers comprise the bulk of the disk space: seventy-three short stories, full novels, and excerpts by writers from thirteen countries, each in the original language. These range from a tenth century verse fragment, through passages by Jonathan Swift (1726) and Voltaire (1752) "describing the two Martian moonlets well over a century before their actual discovery." As well there are fifty-four images of Mars, memorabilia-like movie posters, and introductory audio from luminaries of our era. A story of mine was included, and I helped design the framing of a computer game based on the disk.

I liked the idea of sending a cultural mosaic of how our times envisioned Mars. This seemed more honest than the High Church approach of sending only our best (or perhaps merely snootiest) work. We would depict lurid magazine illustrations beside a Chesley Bonestell painting, *Flash Gordon* along with the classy poetry. Of course there was a Kilroy element, too: inside MAPEX was an electron-beam lithography "dot" carrying the names of all Planetary Society members. This was the first I had heard of a total shotgun approach to a Kilroy message, and in retrospect it seemed to foretell much that followed.

Mars '94 was later delayed to become Mars '96, which failed at launch and splashed into the Pacific Ocean. Both MAPEX and the Planetary Society names did finally make it to Mars on the 1997 *Pathfinder* spacecraft, though, and are now part of the Carl Sagan Memorial Station at the landing site.

The Mars '94 disk brought me into close touch with Jon Lomberg. I was pleased when he came to my campus, the University of California at Irvine, in 1994. He was a visiting U.C. Regent's Lecturer at UCI, a special appointment for nonacademics who have made important contributions.

He had just returned from Carl Sagan's sixtieth birthday celebration, where he gave a talk on "The Visual Presentation of Science," a topic for which he was well-qualified, having collaborated with Sagan on book and media projects since 1972, winning an Emmy Award for his work as chief artist of Sagan's COSMOS series, and later designing the animation sequences for the 1997 Warner Brothers film of Sagan's novel *Contact*. Now in his forties, his paintings adorn many books and exhibitions; his astronomically correct rendering of our galaxy greets visitors to the National Air and Space Museum.

Lomberg immediately brought up an idea he had first ventured in a letter to the Planetary Society in July 1994, and then discussed at the Sagan party: put a message on the *Cassini* spacecraft bound for Saturn in 1997. He had already enlisted the help of Carolyn Porco, a professor of astronomy at the University of Arizona, who promised to get such a message on the spacecraft. Porco was a principal investigator on the *Cassini* imaging camera team, brisk and efficient, and wanted to work on the message itself. In her thirties, she was quick, ambitious, and devoted to her work.

Cassini was to be an anthology mission, with eighteen separate scientific instruments. It also carried a lander that would drop through the soupy atmosphere of Saturn's largest moon, Titan, and radio back data from the surface. A duplicate message might also fly aboard the *Huygens* probe lander (named for the discoverer of Titan), built by the European Space Agency. At 5,562 pounds, the *Cassini* spacecraft would be the heaviest unmanned package ever launched into the solar system, except for the failed Mars '96 craft. With fuel, it weighed 12,470 pounds and was the last of the dinosaur generation of spacecraft, having accreted more experiments as the planning spiraled through many years. Under Daniel Goldin, NASA's approach had reversed to favoring "lighter, faster, cheaper" missions, and Cassini narrowly averted cancellation.

The old-style compendium missions were physically as well as economically vulnerable. The *Galileo* probe had to swing by Venus and Earth several times before reaching Jupiter in 1996, adding years to its journey, because no booster was powerful enough to shoot such a hefty payload to Jupiter in a direct shot. Perhaps due to a swing into the hotter environs of Venus, *Galileo* failed to deploy its main antenna, so the entire mission had to send a trickle of data back on a small antenna never intended to do that job. *Galileo*'s observing agenda had to be cut.

Including staff salaries, and assuming it survives for five operating years in the Saturnian system, Cassini will cost $3.5 billion. It is surely the last multipurpose mission to which teams of scientists glued their hopes and hardware as the mission consumed their careers. Astronomers exploring the outer solar system must deal with long flight times, but the repeated delays of Cassini meant that some of them would have only this single opportunity. After Cassini, missions will be quick, light, cheap—and politically stronger. NASA's extreme sensitivity to Congress came from years of narrowly getting Cassini past their skeptical eyes. The agency grew risk-averse as launch date approached, a fact that came to have great significance as the drama of the Cassini marker unfolded.

While we were working on the Cassini marker, the Mars '96 mission ended up in the Pacific Ocean. It failed to reach orbit because the Russian *Proton* booster misfired in its fourth rocket stage. Again, the craft was so heavy that a fourth stage was essential.

Many experiments were lost, the Visions of Mars disk with them. There was some consolation that the disk may fly on a later Russian Mars mission.

Cassini is an implausibly fat spacecraft, so heavy that it has to undergo two gravity-assist flybys of Venus, and one each of Earth and Jupiter. Arriving at Saturn late in 2004, it will fire an onboard rocket to brake it into the first of some six dozen orbits during its planned four-year tour. Shortly after arrival, the *Huygens* lander will separate and plunge into the cold hazy-brown atmosphere of Titan.

Apparently, Titan has at least one continent, perhaps jutting up from chilly seas of liquid hydrocarbons like ethane. Organically rich, its atmosphere is thicker at the surface than Earth's, but at temperatures around -170 Centigrade. No one has any good idea of what such frigid chemistry could produce, over the four billion years Titan has orbited Saturn.

In November of 1994, Lomberg and I wrote to the Jet Propulsion Laboratory (JPL), which was assembling the spacecraft. As with Mars '94, we suggested attaching an existing small package, the Microelectronics and Photonics Exposure experiment, plus a message. Lomberg thought adding MAPEX might make the marker more salable. We tried to hawk the idea with the usual positives:

- increasing public awareness of the mission, as the Pioneer plaque did, through an optimistic, imaginative goal
- educating a broad public about the lander, Titan's strange chemistry, and the problems of communicating across long timescales

The eventual audience could be humanity centuries hence, or, on a far longer time scale, any life-forms that evolve in the organic soup of Titan. We would not imply that Titan bears life now, but will allow for later evolution. We sketched out the plausible readers, ranging from our distant heirs (1,000 to 100,000 years) to aliens, including possible Titanians, on scales of a million to billions of years.

Perhaps the most important audience would be not distant generations, but ourselves. To many, such messages' true purpose is to appeal to and expand the human spirit. Unlike listening for radio signals from extraterrestrials, sending messages inscribed on hard objects to distant worlds implies a vast time scale before any reward.

Further, such acts demand that we explore what we should say, not how to fathom what others say.

Porco came to UCI and we three spent days in a loose discourse, brainstorming design ideas over lunches and dinners. Much scientific work proceeds like this, sighting in on the critical problems, then using the skills of each team member to attack them. Such free-for-alls are one of the best aspects of scientific collaboration, spirited and enjoyable. They are quite the opposite of how other creative people work, as in the classic image of solitary, agonized artists.

Labor and material costs were to be kept low. We thought the message bearer should probably be an "artificial fossil" embedded in hard glass which could survive Titan's weather. The message would far outlast the lander.

Unlike the wandering Voyager strategy, we could shape our message for a specific place, Saturn and Titan. We could include information about the present solar system (which cannot be seen in visible light through Titan's thick atmosphere) and our place in it. Communicating this in clear, unambiguous ways promised to be an imaginative intellectual exercise, raising interwoven cultural and scientific issues of wide interest. We would aim to be "understandable, optimistic, and awe-inspiring."

JPL said they would submit the idea through the usual channels; Carolyn Porco promised to hurry it along.

Mulling over the huge time scales a week later, I realized that Titan's frigid weathering and the lacerating forces the orbiter would meet around Saturn suggested a message medium of great durability.

Engineers estimated the orbiter would remain intact in orbit for roughly a century, while the *Huygens* lander could be buried by the flows of sluggish, cold fluids within decades. These were very crude projections, given Titan's unknown weather. In both environments, diamond would preserve a message against abrasion better than metals.

To me, the best candidate appeared to be a thin, single-crystal diamond disk to write upon. Using a jewel to carry a message across a billion years could delight the mind, as well.

Manufacturing a disk bigger than a nickel would be expensive. And how to write on the hardest of all substances? At first I thought of using writing processes I knew, such as a layer of boron inside the sheet, laid down using a template and chemical vapor deposition.

The utility of this approach lay in its simplicity, readability, and the unequaled rugged properties of single diamond crystals. Diamond is robust, strong, inert, and resists abrasion. Only very high temperatures and aggressive oxides can damage it. Further, it is transparent in the visible and a broad range of the infrared. Many spacecraft use diamond windows for their infrared sensors, and its space-rated properties are well-known. On Titan, infrared is probably the preferred range for best visibility. Diamond has no known chemical reaction with substances in the Titan atmosphere.

Construction of the marker would begin with purchase of an industrial diamond plate, polished, about one millimeter thick. My discussions with the leading diamond firm, DeBeers, proved this was not a routine request, but they could make such diamond disks for about five thousand dollars each. Since cost scales quickly with size, maximum diameter would be at most a few centimeters.

Writing a microscopic message into the planes of a diamond would probably be attractive to the general audience, I thought, much as the gold-plated Voyager disk proved eye-catching. Indeed, DeBeers seemed interested in the jewelry angle as a possible new market: wear the Cassini Medallion! At perhaps $30,000 or more, this would be a very high-end item.

Lomberg, Porco, and I visited JPL and spoke with the flight engineers and managers, with Porco fielding this proposal in Europe. The jewel message notion seemed to catch the attention of even skeptical engineers. We had approval within a month. The European Space Agency also liked the idea, and agreed to carry a diamond disk on the *Huygens* lander.

Word came to me late in the evening, by telephone from a jubilant Lomberg. I walked outside and viewed the stars, thinking of the marker as a sort of memorial for all the scientific community, and indeed, for our era. The sheer joy of it made it difficult for me to speak. I remembered that awe is a blending of wonder and fear, and realized whence my fear came. The time scales of astronomy imply the mortality of those who study it. No less does designing a message which could not be read until all its designers were dust. The night sky filled me with a chilly awe in a way it never had before.

I went back inside and got to work. Soon enough, consultation with DeBeers converged upon a disk 2.8 centimeters across, a millimeter thick, and weighing 4.3 grams. Each spacecraft would carry the same message. Though we had two years until the diamond disk had to be attached to the spacecraft and lander, there were myriad engineering and conceptual issues to resolve.

TALKING TO ALIENS

In 1862, Victor Hugo had just published *Les Miserables*, and while on holiday wanted to know how it was selling. He sent his publisher a note consisting of a single punctuation mark: *?* Not to be outdone, the publisher replied with *!* This was the shortest correspondence in history, and it is difficult to see how it could be equaled.

This worked because both sides knew enough from context to deduce much meaning from a single sign. Author Tor Norretranders calls this *exformation*: content discarded but referred to by background and circumstances. Exformation can greatly compact messages. Alas, most contexts are present-saturated and quickly pass from the obvious to the unknowable. Who remembers the origin of "23-skiddo," a "hep" expression of seventy-five years ago? The Hugo-publisher correspondence avoided this by using only punctuation. Still, once English has altered or vanished, *?* and *!* will signify nothing except to scholars.

Exformation-rich messages have depth in the sense that they call forth much with few symbols. The more exformation shared by sender and receiver, the more compact a communication can be. The ultimate form is exformation carried by nothing, no information at all. Suppose I agree with you that I won't call by telephone tomorrow if everything is going according to some plan we have. If you hear nothing, you know something, with no signal passing between us. (Unless the telephones don't work, so I can deduce nothing.) Effortlessly, we have achieved the supreme compaction of communication.

Alas, between friends this is simple, but between distant eras and cultures it is nearly impossible. The only reliable exformation is what we share as primates and humans: our way of filtering the world and our innate reactions to it. In working on the Waste Pilot Project, Jon Lomberg and I could at least count on dealing with distant humans. Here, the time and distance scales were so immense that even if humans (not robots or aliens) found the diamond disk, we could not assume much in common with them. They could be the product of a million years of evolution, not the Pilot Project's horizon of a mere ten thousand.

These were our first, daunting thoughts as we contemplated the Cassini message.

We wished to build on the Voyager experience, extending their thinking. As with Voyager, NASA reserved the right to veto us or even drop the marker entirely. When Voyager design ideas leaked to the press in 1977, NASA's official posture was that they had made no final decision on the project at all.

Still, this did not protect from public vitriol the makeshift team making the Voyager record. Shadowy rumors emerged at the United Nations when they tried to get diplomats to record verbal greetings to go on the record. Some felt *Voyager* should carry depictions of war, poverty, and disease, and that a best-foot-forward approach was a sunny half truth.

Early on, the designers had decided to avoid explicit depiction of religion, lest they ignore some. Afterward, others questioned whether the team's belief in the scientific method and use of it to convey much of the message was not itself a sort of ideology. Editorials in the British press had demanded that any future messages be crafted by a large international ecumenical assortment of scientists and nonspecialists alike.

We three had no tolerance for such an unwieldy opera of interests. NASA agreed; we would design and deliver a disk, following solely our own judgment and paying the cost ourselves.

Before beginning, we had to assume that our future readers could indeed read. Brains often must decipher the visual world from ambiguous, ill-defined data. Like many other animals, we make educated guesses about what lies behind our sometimes chaotic environment. Evolution has shaped our brains to create models of the world that mesh well with our learned reality.

At least a third of our approximately hundred thousand genes are exclusively involved in brain function, and many of those relate to sight. We use a strategy of storing a perception across many neurons, much as TV sets break images into pixels.

This method is like the great Rose Bowl prank of 1961, when Caltech students stole the coding sheets for the University of Washington's mass card display. The students then doctored the sheets and returned them to the hotel safe where they were stored. No Washington fan knew the message beforehand, so none could tell that anything was wrong. Each Washington fan knew only to hold up his white or black card, following written orders handed out at the game. When the stadium crowd held aloft their cards, they spelled CALTECH. The next image in this little half-time entertainment was of the Caltech beaver, not the Washington Husky.

Like the fans, our neurons know nothing. But parallel processing of their individual minute signals, carried up through hierarchies of neural organization, eventually constructs a model of what the eye is seeing. The brain uses this image in making evolutionarily effective calculations and decisions.

For example, if we paint dots on a hollow glass cylinder and view it with one eye, it looks like a random set of two-dimensional dots. But rotate the cylinder and—*aha!*—the shape of the glass pops out, a whole three-dimensional picture. Our brain generates this from a mere bit of motion, a talent of great use in the African veldt long ago. Similarly, stereo vision enables our brains to take the small differences in the angles that objects make and decode them into distance estimates.

All this processing plays out behind the sets of our internal, unitary world. We had to assume our future audience would have such abilities as well, but perhaps not exactly ours.

Voyager's messages had embodied the idea that the aesthetic properties of human art (especially music, since they were sending a record) emerged from physical constants and nature's mathematical harmonies. Intelligences of the far future, springing from physical circumstances at least partially shared with us, might well appreciate underlying ideas based on natural order. Lomberg speculated that highly ordered structures like fugues and geometric constructions might come through best.

Conventions of perspective and the entire problem of interpreting two-dimensional representations loomed large. Even those humans whose cultures do not use perspective have to learn how to see it. Dogs never do learn. What of humans evolved in a far future? Or even aliens?

It had always seemed to me that evolutionary mechanisms should select for living forms that respond to nature's underlying simplicities. Of course, it is difficult to know in general just what simple patterns the universe has. In a sense, they may be like Plato's perfect forms, the geometric constructions such as the circle and polygons, which supposedly we see in their abstract perfection with our mind's eye but in the actual world are only approximately realized. Thinking further in like fashion, we can sense simple, elegant ways to viewing dynamical systems, calling forth ideas of the irreducibly elementary.

Imagine a primate ancestor for whom the flight of a stone, thrown after fleeing prey, was a complicated matter, hard to predict. It could try a hunting strategy using stones or even spears, but with limited success, because complicated curves are hard to understand. A cousin who saw in the stone's flight a simple and graceful parabola would have a better chance of predicting where it would fall. The cousin would eat more often and presumably reproduce more as well. Neural wiring could reinforce this behavior by instilling a sense of genuine pleasure at the sight of an artful parabola.

We descend from that appreciative cousin. Baseball outfielders learn to sense a ball's deviations from its parabolic descent, due to air friction and wind, because they are building on mental processing machinery finely tuned to the problem. Other appreciations of natural geometric ordering could emerge from hunting maneuvers on flat plains, from the clever design of simple tools, and the like. We all share an appreciation for the beauty of simplicity and the ugliness of conceptual clutter, a sense emerging from our origins.

In an academic paper, Guillermo Lemarchand and Jon Lomberg had argued in detail that symmetries and other aesthetic principles should be truly universal, because they arise from fundamental physical properties. Many earlier messages and ideas about radio communication had made similar assumptions. Aliens orbiting distant stars will still spring from evolutionary forces that reward a deep, automatic understanding of the laws of mechanics.

Many things in nature, inanimate and living, show bilateral, radial, concentric, and other mathematically based symmetries. Our rectangular houses, football fields, and books spring from engineering constraints, their beauty arising from necessity. We appreciate the curve of a suspension bridge, intuitively sensing the urgencies of gravity and tension.

Radial symmetry appears in the mandala patterns of almost every human culture, from Gothic stoneworks to Chinese rugs. Perhaps they echo the sun's glare flattened into two dimensions.

In all cultures, small flaws in strict symmetries express artful creativity. As Lemarchand and Lomberg note, the universe itself began with a Big Bang that can be envisioned as a four-dimensional symmetric expansion; yet "without some flaws, so-called anisotropies, in the symmetry of the Big Bang, galaxies and stars would never have appeared."

A less obvious mathematical underpinning expresses itself in forms as diverse as the chambered nautilus, flower petals, and galaxies. Draw three diagonals in a pentagon, and the intersections divide the lines in a ratio, $1/2(1+5^{1/2}) = 1.61803 \ldots$ The ancient Greeks noticed that this "Golden Section" in geometry emerged in many strikingly different ways. The human eye finds it pleasing in our own buildings; the Greeks knew this.

When its pediment was intact, the Parthenon fit exactly into a rectangle with this ratio of sides. The proportion was first discovered by the Greek mathematician Pythagoras 2,500 years ago; the sculptor Phidias used it. The United Nations building in New York City is proportioned as three stacked Parthenons.

Natural philosophers noticed that this number also appears in a famous sequence, the Fibonacci series (0, 1, 1, 2, 3, 5, 8, 13, 21 . . .), which nature favors as well. Arrived at simply by summing the previous two entries in the sequence, this pattern appears in the branching of trees, in the number of petals in the iris, primrose, and daisy, and in many other flowers. Pinecones, pineapples, and sunflowers display overlapping clockwise and counterclockwise patterns, their florets in the ratio of successive Fibonacci numbers, such as 21:34 in the sunflower. The Golden Section emerges when one takes the ratio of two successive terms; the higher these terms are, the nearer their ratio to 1.61803 . . .

The Golden Section emerges from spirals by drawing perpendicular lines connecting different parts of the curve. The ratio of the lengths of adjacent lines is a close approximation to 1.61803 . . . The spiral of the chambered nautilus follows the Golden Section, as do the curves of seashells and animal horns. Apparently, the necessities of strong structures built from minimal materials force such underlying choices to emerge from the pressures of evolution. Growing in a fixed proportion does not shift the center of gravity, so balance problems do not develop.

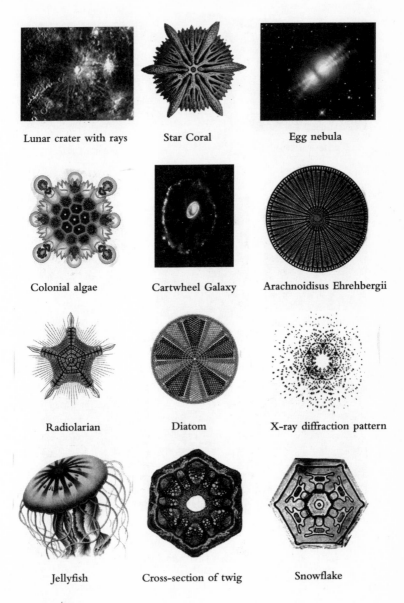

Lunar crater with rays	Star Coral	Egg nebula
Colonial algae	Cartwheel Galaxy	Arachnoidisus Ehrehbergii
Radiolarian	Diatom	X-ray diffraction pattern
Jellyfish	Cross-section of twig	Snowflake

Figure 2.1 Radial symmetries in nature (courtesy of Jon Lomberg)

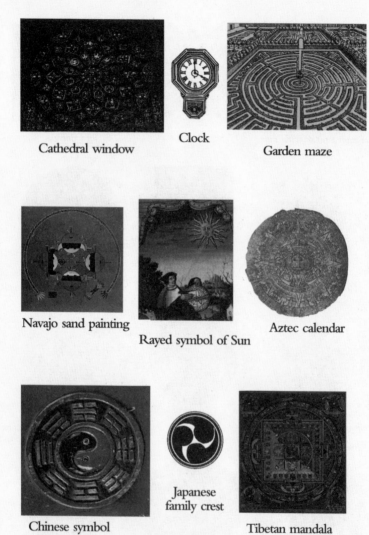

Cathedral window

Clock

Garden maze

Navajo sand painting

Rayed symbol of Sun

Aztec calendar

Chinese symbol

Japanese family crest

Tibetan mandala

Figure 2.2 Radial symmetries in art (courtesy of Jon Lomberg)

Chambered Nautilus
(Nautilus pompilius)

Galaxy M74 in Pisces

Sneezewort (Achillea ptarmica)

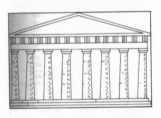

The Parthenon at Athens, was built in the 5th century B.C. When its triangular pediment was intact, its dimensions could be fitted almost exactly into a Golden rectangle, as shown above.

The Parthenon at Athens, was built in the 5th century B.C. When its triangular pediment was intact, its dimensions could be fitted almost exactly into a Golden rectangle, as shown above.

A different connection with Fibonacci numbers is found in the number of axils on the stem of a plant as it develops. An ideally simple case is represented in the figure, where the stems and flowers of sneezewort are set out schematically. A new branch is seen to spring from the axil and more branches grow from the new branch. Since the old and new branches are added together, a Fibonacci number is found in *each horizontal plane*.

The golden numbers come into view again if we examine the number of petals of certain common flowers; examples are:

Iris	3 petals
Primrose	5 petals
Ragwort	13 petals
Daisy	34 petals
Michaelmas daisy	55 and 89 petals

Figure 2.3 The Golden Section in nature and art (courtesy of Jon Lomberg)

ϕ and Fibonacci numbers

If $u_3 = u_2 + u_1$, and $u_1 = 1$, and $u_2 = 1$, then we derive the following series:

1,1,2,3,5,8,13,21,34,55,89,144, etc.

This is called the Fibonacci series.

The Golden Ratio, ϕ may be expressed as $\phi \to u_{n+1}/u_n$ as $n \to \infty$

For u_{30}/u_{29}, $\phi = 1.6180339887498744831$

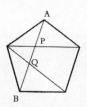

The point of intersection P of two diagonals of a pentagon divides each in the golden ratio *phi*. P divides AQ and AB internally and QB externally in this ratio.

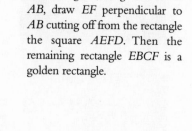

Multiple reflections of light rays between sheets of glass are Fibonacci numbers

In *ABCD*, in which $AB:BC = \o:1$; through E, the golden cut of AB, draw EF perpendicular to AB cutting off from the rectangle the square $AEFD$. Then the remaining rectangle $EBCF$ is a golden rectangle.

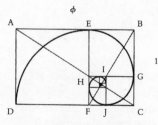

Logarithmic spiral

If from this the square $EBGH$ is lopped off, the remaining figure $HGCF$ is also a golden rectangle. We may suppose this process to be repeated indefinitely until the limiting rectangle O, indistinguishable from a point, is reached.

Figure 2.4 The Golden Section in mathematics and physics (courtesy of Jon Lomberg)

Quite different physics generates the spiral waves in galaxies, yet in many these curves, too, express the Golden Section, sometimes also called the logarithmic spiral. The Golden Section lives in flowers, trees, and galaxies, giving pattern to our entire universe, yet known only to a few of us hominids.

To those who have not had their sense of mathematics squashed by the mechanical drills of elementary school, the subject can burn with a peculiar rich intensity. Would aliens (or even further evolved humans) "see" the same mathematical underpinnings to our universe?

Strategies for the Search for Extra-Terrestrial Intelligence, SETI, have assumed this since their beginnings in the early 1960s. Many supposed that interesting properties such as the prime numbers, which do not appear in nature, would figure in schemes to send messages by radio. A case for the universality of mathematics is in turn an argument for the universality of aesthetic principles: evolution would shape all of us to the general contours of physical reality. The specifics could differ enormously, of course, as a glance at the odd creatures in our fossil record shows.

Our prospect was daunting. Many mathematical paths beckoned. For example, was there a way to embed in our message the compact equation $e^{i\pi} + 1 = 0$, which links the most important constants in the whole of mathematical analysis, O, 1, e, pi, and i? The equation looked beautiful to me—a "math type," as my wife dryly noted—but such types comprise a tiny audience even among humans.

What's more, we could not find a clear way (independent of many assumptions about notation) to even write the equation. Any writing scheme called upon human symbols. Such points stumped us. After all, philosophers of mathematics have argued over whether a mathematical object, like 9, is independent of human thought, or not. Some hold that it is neither external nor internal but social. This means mathematical ideas arise from our interactions with each other. Then a theorem known solely to its inventor does not in some sense even *exist* as mathematics until someone else understands it. Plates are round, an objective fact, but mathematical roundness is a human construction.

Perhaps. But all three views—mathematics is objective and real; it arises from an internal set of preconceptions; it is social—ignore biology, which brought about humans themselves through evolution. How general were our adaptations to our world?

How to decide such fundamental points? Our imaginations yearned to soar but momentarily stalled. In the end we retreated to our sense of beauty.

Further difficulties arose in areas I had naively thought were straightforward. How to depict our solar system? To use mathematical universals, even once identified? How about the data processing assumptions behind recovering three-dimensionality through two-dimensional projections? How universal could be the use of scientific diagrams, our design of mathematical symbols, and the use of photos of humans?

All involved standing at a conceptual distance from ourselves and reaching for a more general way of seeing the world. But how firmly could we believe arguments from our own sense of beauty?

IMAGINING OTHER MINDS

Our attempts to communicate with other species on Earth have not been strikingly successful. We have developed a dictionary of several hundred words for communicating with bright chimpanzees. Very limited discourse flows between us and the cetaceans, principally dolphins, though whale songs are intriguingly complex; more research in these directions would be illuminating.

But these are minds evolved in our own biosphere. Surely contemplating a message that utterly different minds could fathom sets the outer limit of any deep time strategy. Even if we look for mathematical universals, how truly general are they? Here, conversations Lomberg, Porco, and I had with the noted theoretical psychologist Louis Narens proved unsettling.

Semiotics is the study of nonverbal communication, and mathematics is fundamentally nonverbal, though ours (and presumably others') does have a notation. Scientists usually think of mathematics as a set of immutable truths, emerging from the world with the heft and solidity of Platonic ideals.

Narens dispelled these comfortable notions. Alien arithmetic could be nonnumerical, he said—that is, purely comparative rather than quantitative. Such alien beings might think solely in terms of whether A was bigger than B, without bothering to break A and B into countable fragments.

How could this arise? Suppose their surroundings had few solid objects or stable structures—say, they are jelly creatures awash in a thick Titanian ethane sea of the far future. Indeed, if they were large creatures requiring a lot of ocean to support their grazing on lesser beasts, they might seldom even meet each other. Seeing smaller fish as mere uncountable swarms—but knowing intuitively which knot of delicious stuff is bigger than the others—they might never evolve the notion of large numbers at all. (This idea isn't crazy even for humans. The artificial intelligence researcher Marvin Minsky told me of a patient he had once seen who could count only up to three. She could not envision six as anything other than two threes.)

For these beings, geometry would be largely topological, reflecting their concern with overall sensed structure rather than with size, shape, or measurement, à la Euclid. Such sea beasts would lack combustion and crystallography, but would begin their science with a deep intuition of fluid mechanics, as obvious as gravity is to us.

Of course, these creatures might never build any technology, and so not find our diamond marker in the sludge of Titan, much less the one on the *Orbiter* around Saturn. Even land-based creatures might not share our assumptions about what's obvious. Our concepts are unsuited to scales of size far removed from those of our everyday experience. What might Aristotle have thought of issues in quantum electrodynamics? He would have held no views, because the subject lies beyond his conceptual grasp. His natural world didn't have quanta or atoms or light waves in it. In that very limited conceptual sense, Aristotle was alien.

Our hope for the message was that at least, and perhaps only, in the cool corridors of mathematics could there be genuinely translatable ideas. Marvin Minsky takes this view, believing that any evolved creature—maybe even intelligent whorls of magnetic field, or plasma beings doing their crimson mad dances in the hearts of stars—would have to dream up certain ideas, or else make no progress in surviving, in mathematics, or in anything else. He labels these ideas Objects, Causes, and Goals.

Are these fundamental notions that any alien must confront and use? We cast a pale shadow of doubt over Objects, since these depend upon one's perceiving apparatus; a snake would respond to images in the infrared while we would not, for example. Even causality isn't a crystal-clear notion in our own science, particularly in quantum mechanics.

Why, then, should Objects, Causes, and Goals emerge in some otherworldly biosphere? Minsky holds that the ideas of arithmetic and of causal reasoning will emerge eventually because every biosphere is limited. Some inevitable scarcity will occur. Limited means force adoption of realizable Goals, and reward those who Caused them to come to pass, acting upon Objects.

Organisms that can conceptually organize the external world to perceive Minsky's root ideas will get more from their efforts. Such Darwinian selection will affect all their later biases. Minsky has framed technical arguments showing that these notions must turn up in any efficient (and, presumably, intelligent) computer, and the ideas may generalize to aliens. Of course, there is a big conceptual leap here. Computational ideas may not prove adequate for describing biological organisms; they certainly weren't obvious even to us a mere century ago.

Most scientists who have thought about communication with aliens work within the assumptions of SETI—with mathematics as the fulcrum of communication. Hans Freudenthal's LINCOS is a computer language designed to isolate the deepest ideas in logic itself and to build a language around it. It uses binary symbols typed out in lines (a choice we also made for our message). LINCOS stands ready the moment we run into something green, slimy, and repulsive, and yet with that restless urge to write—or read.

Math is central to the whole issue of communication because it allows us to describe "things" accurately and even beautifully without even knowing what they are. Richard Feynman once said, to the horror of some, that "the glory of mathematics is that we do not have to say *what we are talking about*." (emphasis his) Feynman meant that the "stuff" that communicates fields, for example, will work whether we call it wave or particle or thingamabob. We need not have such cozy pictures at all, as long as we write down the right equations. As David Politzer of Caltech once remarked, "English is just what we use to fill in between the equations."

We felt this might help as we sought what Louis Narens termed "cognitive universals." While it is impossible to avoid biases because we are humans, and immersed in our cultures, we had also to avoid regarding some ideas as obvious simply because we could not imagine how they could not be. As Minsky sardonically remarked, "Artificial realms like mathematics and theology are built from the start to be devoid of interesting inconsistency." The real world could summon forth fantastic choices.

Narens had made a vital point: while aliens should have concepts like arithmetic, they need not have more than rules for how to add, subtract, and multiply *particular* numbers. We reason inductively, as in the title of a George Gamow book, *One, Two, Three . . . Infinity*. We humans generalize simple, small numbers to arbitrarily large ones, and invent relations between them as well.

This gives us a fascination with primes, for example, and how to calculate large ones, as part of an infinite ensemble. To us, numbers are Platonic objects, existing as ideals, literally innumerable in extent. Aliens may not need such a habit of mind.

The abilities necessary for generalizing based on induction seem to come from the linguistic abilities of creatures on Earth, particularly us. Hunters in the animal kingdom have counting ability, as would come naturally from a sharp sense of how to cut an ailing member from a herd—that is, abstracting an integer from an unbroken flow. But evolutionary pressure for more efficient processing need not necessarily lead to inductive generalizations like the totality of natural numbers. Few people, after all, need to use numbers larger than a hundred or so except in financial matters.

Instead of concentrating upon general facets of integers, for example, one could imagine minds that "see" the flux of a physical quantity. Indeed, for some important features of our world, we, too, perceive intuitively the flow of things, not the quantity itself. We speak of a room being chilly, but we do not measure its temperature; instead we sense the rate that heat flows out of us, a derivative of temperature. When we think of metal being colder than wood, for example, we are actually discerning our different heat losses through them. Evolution has geared us warm-bloods to be leery of losing energy.

Similarly, our vision registers the logarithm of light intensity, not the intensity itself; this is why we can see over such a wide range of brightness, from noonday glare to starlight.

Viewing our problem so broadly was invigorating but daunting. Imagining weirdly different readers of our diamond disk was entertaining, but it gave us too much latitude. How could we ever decide on specifics?

In the end we fell back upon the aesthetics argued for earlier by

Lomberg. Still, we had to remember our earthly audience; a bizarre message might play poorly in Peoria. Narens agreed, remarking that "bad presentation of the rationale for such messages, their design, or their content, could easily generate ridicule—not only for the particular message but also for creation of such messages in general. A good job could add another dimension of adventure to the mission."

To impart numbers, we decided to use binary notation, following LINCOS. This was an almost inevitable choice, since digital portrayal emerges naturally, mandated by the fact that any number enjoys a unique representation only in base 2. The number of days in Earth's year is:

$$365 = 2^8 + 2^6 + 2^5 + 2^3 + 2^2 + 2^0$$

Only in base 2 is this designation, 101101101, unique. Thus communication between any entities who fathomed mathematics, and understood integers, could well be forced to the common tongue of binary notation.

Why is binary so basic? Because plus/minus, greater or lesser, up/down are simple distinctions useful in any environment. Life just about anywhere would have to make such contrasts.

We could cement this by expressing the Golden Section, $\frac{1}{2}(1+5^{1/2})$ in binary. But how to decide where to cut off the infinite fraction, which does not repeat? This issue we never settled.

We decided to employ a mass and length standard related to the spacecraft, just like the Voyager "dictionary" prepared by Frank Drake and Lomberg. Depiction and encoding would follow Voyager methods. Since the diamond might well survive alone long after the rest of the spacecraft disintegrated, its diameter of 28mm was the obvious unit length, which we termed "1u." Saturn's mass we gave in units of the disk mass, 4.32 grams, so the number in binary stands for 1.3×10^{29}. To make the connection, we placed it beside a picture of Saturn itself.

What else was truly basic? As a physicist I thought of unit systems, but other sciences have other fundamentals. Mark Martin, an Occidental College biologist, suggested including a compact two-dimensional sketch of our biosphere. Accompanying the planned picture of humans, it could show the major kingdoms of life, with size markers beside the simple line drawings to give their true scale.

This would presume aliens could decode such a 2D picture; many animals, after all, cannot.

I then thought of including a truly dense carrier of information, a strand of DNA. This basic "unit" of biology would adroitly send much biological knowledge, independent of language, by giving our readers the thing itself, not a representation.

Of course, such a DNA record is not useful without the record "player": specifications of intricate conditions in the womb, etc. But Lomberg quickly pointed out that to send any biological sample into space, however tiny, would violate the Planetary Protection Protocol, a set of rules to stop contamination of worlds. Never mind that the DNA could be sealed inside the diamond disk; jumping through bureaucratic hoops would take time and effort we did not have.

He even raised a moral issue: what if someone in the far future "resurrected" the being from its DNA? Should we "condemn" a genetic descendant to whatever world it would face? This struck me as a bizarre argument, but it had the smell of Voyager's reception, too: someone would undoubtedly raise this, getting plenty of splashy coverage. ("NASA Sends Elvis Sperm to Saturn," Lomberg suggested all too plausibly.)

Given these troubles, we rejected any biological sample.

But what of less obvious standards? Time has no clear units appropriate to billion-year scales; atomic frequencies are in the billion*th* of a second. Here we decided to rely on a crucial assumption: that whoever found the message would know a bit of astronomy, even if they lived in Titan's soupy atmosphere.

MARKING TIME

If in some far distant future other beings were to find the Cassini/Huygens Diamond Medallion, we assumed they would wonder who made it, and when, and what it has to tell them. Surely beings who reach orbit about Saturn should have some trait resembling what we call curiosity.

To answer such questions we designed several two-dimensional pictures. Could we trust that our future audience could comprehend flat, two-dimensional images? Again, invoking evolution, we gambled that intelligence would arise in beings who had followed strategies based on clever gathering, if not hunting. Such minds would need vision that could distinguish distance.

Of course, we are prejudiced in favor of the visible spectrum. Bats, whales, and dolphins see with acoustic waves. We quickly decided to give up any hope of communicating with such creatures, unless they could sense the patterns we would etch into the diamond by reflected sound waves. This seemed quite unlikely, given the tiny writing we were forced to use. Acoustic waves of that size are very high frequency.

Further, quite plausibly our audience would know that objects that in a flat projection are above others would most likely be further away, just as in a landscape painting the horizon is higher than nearby rocks. Vision would have to deal with troubles such as, say, trees that begin at the ground and cross the horizon. This need not require binocular vision, but we did demand some such ordering principle of our audience. Without it, our choices were simply too broad, and might elude our Earthly audience as well.

Photographs of the Earth should show where the message came from, and a stereo photograph of humans will show them who made the diamond. As well, the stereo photo could help show how we turn two-dimensional representations into three-dimensional ones.

But how to show *when* the diamond was created? A calendar based on our present reckoning would mean nothing.

Previous deep time messages used astronomical objects as time markers. In the Voyager Interstellar Record, Frank Drake suggested including a photograph of the Andromeda Galaxy, our own Milky Way Galaxy's nearest large neighbor. Andromeda is visible to our naked eyes, and should be obvious from anywhere in the Milky Way. Galaxies rotate in about half a billion years, and several dwarf galaxies orbiting Andromeda move perceptibly in a million years. Those who find the Voyager Record in interstellar space millions of years from now can compare how Andromeda looks then with our photograph. If they are good astronomers, they can then estimate *Voyager*'s age to within perhaps a million years.

In a similar sense, photographs of the Earth also provide a kind of calendar, since continental drift is perceptible on time scales of a few million years. Continent profiles and positions on the future Earth, compared with an Earth map that was part of the message, could lead an astute alien to a rough age, assuming they understand continental drift and could see Earth at all. Present Titan residents could not make out the inner planets in visible light through their thick atmosphere. Also over many hundreds of millions of years the continents will alter unrecognizably, and the diamond's readers may also not have good close-up images of our planet.

This strategy was used already, on a small plaque attached to the LEGEOS satellite. Carl Sagan designed a picture of our continents as they looked several hundred million years ago, their present map, and a projection of how we think they will appear in eight million years, the rough lifetime of the satellite.

With this in mind, Lomberg proposed using four different types of astronomical objects that evolve at different rates, "clocks" covering different time scales. Whenever the diamond is found and studied, one of the photographs will display changes allowing a rough dating.

SATURN

Anyone who finds the diamond in orbit around Saturn or on Titan will know the looming presence of the nearest planet. Saturn's beautiful ring system can be our clock. Saturn's tidal pull prevents these disks of ice particles from collapsing into a moon. The rings are extremely complex, like giant phonograph records, with thousands of ringlets separated by small gaps. There are large gaps as well, the most visible from Earth being the Cassini Division. The French-Italian astronomer Giovanni Domenico Cassini discovered the gap in 1675, correctly believing the rings to be made of particles, though this view took a century and a half to find acceptance.

At the Cassini Division, particles orbit Saturn in half the period of the innermost moon, Mimas. Mimas exerts a cumulative tidal pull on these, tugging them out of the gap region. The division's exact position then depends on Mimas's orbit, which itself slowly changes over hundreds of millennia, due to the tidal forces of Saturn's other moons. As Mimas moves, so moves the Cassini Division.

If those who find the diamond know this, careful comparison of the division's precise position as they see it with the diamond's picture will serve to date the diamond. This calendar should be useful over periods of millions of years. The limitations of the calendar lie in our own knowledge, for we have no sure idea of how long Saturn's rings will last; they may be a passing phenomenon, on astronomical scales.

THE BIG DIPPER

Constellations are chance cliques of stars, unassociated except for being grouped in our night sky. Each star moves through the galaxy in its idiosyncratic orbit, so the accidental association into, say, the Big Dipper will disperse as the stars move on in different directions at different speeds. Our sun moves, too, and the sum of all these motions ultimately will take millennia to alter a constellation in our sky.

Constellations look the same from Saturn as from Earth, because the stars are so far away. Our famous and easily recognized Big Dipper will slowly alter over tens of millennia. Its bowl will become shallower as the four stars forming it disperse on their own orbits. Within sixty thousand years distortion will make it unrecognizable. The Waste Interment Pilot Project panel had proposed this same time marker, developed by Drake, Ben Finney and Lomberg.

From the Big Dipper's photo on the medallion, its discoverers should recognize it but also notice its altered shape. If they understand astronomy, they could compare the Dipper as they see it with our image. Within roughly a 20,000 to 50,000 year span, this will give them the time elapsed since it was sent from Earth.

GALAXY M74

Galaxies are bee swarms of stars. Many of them resemble our spiral-shaped disk. Such spiral arms draw the eye to young stars. The curves trace bright star-births that flare in the wake of compression waves. When we see the gaudy, beautiful arms that seem to wrap around the central hub, we are witnessing where dark clouds once collapsed, driven by the characteristic soundlike waves that act upon the galaxy's "gas" of stars and dust. In a few million more years the wave will have advanced a bit, illuminating a different sector.

Galaxy M74 is a spiral that will change similarly to our own. Again, discoverers of the diamond disk can use its M74 photograph as a calendar by comparing the spiral's appearance then and now, discernible over periods of ten million years.

The diamond marker's disk may suggest studying the depicted

spiral disk, but we intended a more basic, elegant subtlety: the Golden Section. To further call this relation forth, we planned to include in the foreground of the photograph of humans, a chambered nautilus seashell, exhibiting the same spiral. We felt that patterns repeated elsewhere on the disk would call attention to themselves.

Spirals appear in seashells and galaxies alike, calling forth our appreciation of beauty. Among the many spiral galaxies, we chose M74 as one of the most beautiful to human eyes. We cannot know if extraterrestrials will share our same sense of the sublime, but we can suggest our own.

As a time marker other galaxies could work better, such as M51, which has a small satellite galaxy orbiting it. The satellite might be easier to track than the spiral arms, its position serving as a clock. On the other hand, M51's spiral arms do not give as close a fix on the Golden Section as M74. This is because galactic arms have thicknesses, so fitting them to an exact line introduces ambiguity. In the end, M74's closeness to the ratio won out over the satellite, a judgment call.

THE HERCULES CLUSTER OF GALAXIES

Galaxies cluster; our own Milky Way Galaxy and neighbor Andromeda are the largest members of a small throng of about twenty galaxies, mostly dwarfs, known as the Local Group. Many clusters of galaxies are much larger, with hundreds or thousands of members. The diamond medallion's photograph shows a section of a large, obvious, nearby swarm, the Hercules Cluster, where many galaxies orbit in a grand gavotte precisely played by gravity. Two such galaxies are colliding. As they separate, they will draw streams of stars out in long tails, easily seen by astronomers with telescopes no more advanced than ours of fifty years ago.

Their slow dance takes the galaxies tens, hundreds, or even thousands of millions of years to alter their positions relative to each other. The Hercules Cluster therefore serves as the longest of our time markers, giving recipients of the Cassini/Huygens medallion a clock ticking off billions of years.

On such time scales, mere mortal beings seem nothing. Yet such beings sent *Cassini*, and presumably other mortals would want to know a lot about their own kind. As with *Voyager*, we wished to send a talisman for all of us.

A PORTRAIT OF HUMANITY

Voyager slipped into interplanetary space, and soon after, the interstellar realm, bearing a large phonograph record with images and sounds of Earth: a portrait of humanity. We wished to do something similar on a tiny surface, inscribing a photograph into the diamond. Discussions ranged over many choices; I shall try to convey the flavor, often quoting or paraphrasing.

Based on the Pioneer and Voyager experience, Lomberg drew up a list of requirements that would both deflect the sort of criticism Pioneer's sketch of humans aroused, and minimize confusions among the eventual readers: "The photo must work in black and white and at low resolution, showing a representative sample of humans with regard to age, sex, coloring, ethnic type, body type, dress, and hairstyle." This last was to show that hair was natural and variable, yet was not clothing, which would also be deliberately diverse.

Further, the photograph "should show the entire human body, from head to toe, in several different positions" to give an idea of the range of movement. With "a minimum of overlap of detail in the poses—that is, people not partially obscured by others"—all objects would clearly stand out from background, and all individuals have equal visual importance. There would be social implications read into the photo by us and by any future readers, but at least we tried not to send signals we did not intend.

All felt that the picture must represent the planet without being too specific, certainly not a unique site or climate. The background should be information-dense, rich with details about the planet, species, and culture, though without compromising any of the above goals. Among this minutiae should stand out an object identical to something on the spacecraft to provide an unambiguous check of scale.

To stand for all humanity, we felt the photo should be open to humanity's inspection. Early on we agreed there would be no permission, or future reproduction, limitations.

Most of all, the photograph should convey beauty and wonder

to our human eyes. We wanted no clinical examination of the human body, but an evocation of ourselves immersed in our world—and Cassini coming out of that passion.

Since we had discussed these ideas while walking in Laguna Beach, we immediately thought of an ocean setting. Lomberg lives on the Big Island of Hawaii, where isolated sandy beaches boast steady conditions for photography. With submerged lava rocks and waves rolling in, clouds in the sky, flapping birds and possibly a waning gibbous moon, we could convey much to a single glance from alien eyes. The beach should not be well-known or easily identified; we wanted a generic beach that could be found on most continents or islands of Earth. Beaches have strong mythic and biological associations that enhance their relevance.

The photographer whom Lomberg knew best and wanted to work with was Simon Bell of Toronto. Bell is one of the world's best stereo photographers, and kept the team from error many times. For his convenience, Lomberg and Porco considered a lake beach shot on the sandy shores of Lake Ontario. In the end this idea lost out. A lake does not impart the same feel as the ocean; waves are smaller and typically the coast is less varied.

Some facets we sought to convey more subtly: "the use and role of boats; the importance of water; the nurturing of children; information about the water cycle and thus the approximate temperature." Cast shadows might imply the latitude or time of day, but for distant eyes that would be difficult.

We see the world in stereo, and a direct way to convey this would be to etch two stereo views of the same scene. This would also strongly hint that the curious bipeds in the photo saw with the odd symmetric spots on their upper heads. The disk's size serves to fix uniquely the distance between the "eyes" of the stereo views, so that images at all distances align properly. The camera separation for the shots was close to our own eye separation, again suggesting that's what our eyes are for.

The one object from the spacecraft we knew future viewers would have was, of course, the disk. One of the photographed adults should then hold the diamond disk, very clearly outlined against a background. To help alien perceptions, all or most of the people should be looking at it.

Porco reminded us of the Pioneer drawing, which to some implied a man was larger and thus more important than a woman.

Never mind that the Pioneer team carefully used figures with the average height of men and women worldwide, they drew objections. Porco insisted that we should make a woman the focal point of the photograph, and Lomberg agreed. When finished, the photo provoked one woman to comment, "It'll tell them in the far future that Earth is a matriarchy. I love it!"

In the photo (Figure 2.5, opposite) the central, seated woman holds the disk (actually, we didn't risk using the diamond, so substituted a plexiglass stand-in with the same optical reflecting characteristics). Others look at it. Its diameter sets the scale of the people and plants within view.

We agreed that while the adults and older child should be clothed, to avoid the Pioneer criticism, the younger children might be nude, to hint at how we reproduce. Casual, loose, and solid-colored, clothing should be shot to make it as easy as possible to see that it is a covering, and not a growth of the body. Women should wear little makeup, if any. Some small jewelry like rings or bracelets might be all right, if it were obviously artificial. No cross or other religious symbols, though; no favoritism should be implied.

Some people we spoke with thought it dubious to not show all people nude, for clarity. But many would object to or be embarrassed by pictures of naked adults. Lomberg carried the day by saying firmly, "If we want this photo to truly be representative of all the Earth, it is no small matter to alienate a large portion of the Earthly audience."

Also, Lomberg noted, people hardly ever walk around naked. In most cultures there is some sort of dress, a fundamental social fact about us. Shadows on the ground and a sun hat could give the very important information that we cover ourselves for protection from the environment. Astute observers might even draw some conclusions about avoiding too much solar ultraviolet at the beach.

As well, sexual differentiation will be guessable purely by obvious body shape differences and the breasts of the women. (But would nonmammals guess their use?) If these hints proved insufficient, the genitals would not provide strong clues as to their function. Necessarily there would be unseen parts of the body—soles of the feet, inside of the mouth—so we could not be utterly, clinically representative. The picture was not aiming to explain human biology

Figure 2.5 The Cassini portrait of humanity (courtesy of and copyright 1997 by Simon Bell and Jon Lomberg)

or reproduction fully, but to satisfy the simple question: What did the creators of this message look like?

The background could explain larger aspects. A shot angled along the beach would show incoming wave trains clearly. Iron-wood conifers along the shore would include another great king-dom of life; the bacteria we would have to do without.

A collection of several different boats—wooden canoe, modern sailboat, fishing boat with motor—might suggest our range of tech-nology and our interest in traveling and vehicles, of which *Cassini* is one of the ultimate expressions. But more than one would clutter the composition, too. Sail size and mast height the viewers could use to roughly estimate our wind speeds and atmospheric density. Birds in the sky would ring in another animal life-form, but how could we count on them? A trained parrot balanced on a limb? This proved difficult to bring about during the long, grueling photo sessions. And a sitting bird would not imply flight. Luck would have to give us that, then, from the myriad shots necessary to get just the right one.

The time of day was another variable, but we could not see how to use it to carry much information. The sun's angle should be low enough to cast clear shadows but not so low as to cause problems with exposure times. The fidelity of the process that would inscribe the photo also set limits. In low-resolution black and white with little dynamic range, our main goal was clarity and clear outlines of objects, though stereo images and foregrounding important ob-jects would help in sorting it all out, we hoped.

Naturally, we thought of the most dramatic possible shot, a sun-set over the ocean. But Simon Bell shook his head. "Using flash may not work. Because I use two cameras, I have to slow the shutter speed to ensure that the flash is caught by both cameras. This would then affect the look of the waves, which we'd ideally want to freeze with a fast shutter speed."

Sunset also drastically reduces potential shooting times and lo-cations and might compromise esthetics, since a good sunset looks best when the sky and foreground are underexposed, while skin tones look best when normally exposed. Using too much fill light to compensate could look artificial, too.

Myriad such considerations entered in the final, four-day-long photo session, at two different beaches. The logistics proved almost military in scope, down to camping gear, food supplies, and a portable toilet. Lomberg organized all this, selected the sites and found the multiracial models, all residents of Kona. Simon Bell flew in from Toronto after trying model poses in his studio, and shot over twelve hundred slides. Some had full frontal nudity, others were unclothed but more discreet. Some were fully clothed, as a hedge against NASA's suddenly balking at the last minute. Porco took the final candidate slides to Washington and showed them to NASA administrator Dan Goldin, who approved the final selection.

As luck would have it, the best shots had no birds flapping in the deep blue sky.

AN ENCYCLOPEDIA ON A PIN

Now came time to assemble all this into a final design.

We planned to inscribe the reverse of the diamond with a straight line across the diameter, broken in the middle to show a 1u symbol. The *u* stood for "unit" and was repeated elsewhere to show we were thinking of lengths.

Given sufficient dynamic range, the refractive characteristics of the diamond might be measurable in the photo. This might be used by readers to reconstruct some of the color. Bell doubted that our sixty-four gray levels would be up to this, though.

Above the 1u line we depicted the first ten digits in binary code, then used binary everywhere else. In a sense, the reverse side sets the terms of discussion: the photo says "This is who we are," while the units tell them how we shall describe our world on the front side. Only the names of supporting laboratories and organizations at the very bottom carries any Kilroy aspect.

From top to bottom the scale goes from small to large. So also on the front side, with the Big Dipper, M74, and the Hercules Cluster at the bottom and the Earth at the top.

Above the stellar pictures is an accurate, scaled map of the solar system, showing planets, symbols for them, and the *Cassini-Huygens* trajectory through them, establishing our home. Above this is a gallery of previous spacecraft such as *Viking*, *Pioneer*, and *Voyager*. (This actually violated the larger-to-smaller scheme, alas, but seemed necessary to tie the idea of such vessels to the planets they explored, shown in a matching gallery of planet photos directly

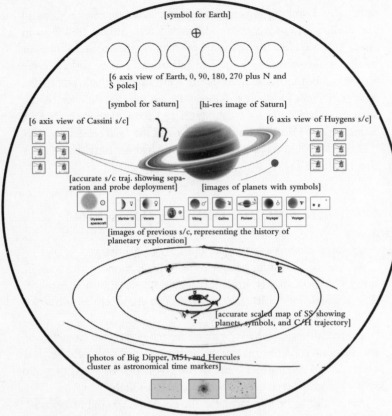

Figure 2.6 Astronomical side of the Diamond Disk design, January 1997 (courtesy of Jon Lomberg)

above each spacecraft. Only Earth has no craft, suggesting that we live there.)

Above this gallery is a highly detailed Saturn with its planetary symbol. An accurate trajectory marks the *Cassini* orbiter separation from the *Huygens* probe. To the left is a six-axis view of the *Orbiter*, to the right of *Huygens*. Above this row are Earth photographs, four of ninety-degree rotations and two of the poles. Tom Van Sandt of the Geosphere Corporation generously gave us cloud-free, consistent lighting maps. From these we hoped continental drift dating could be done.

Many compromises lie behind these views. Could eyes in the far future translate the slanted perspective of the solar system diagram into a three-dimensional reality? We could but hope.

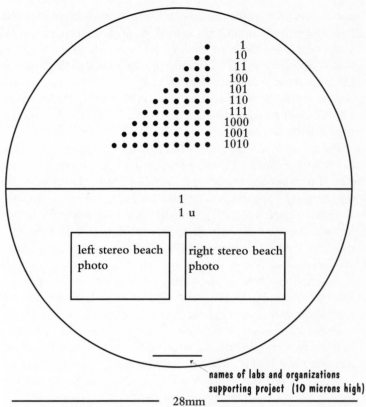

Figure 2.7 Symbolic side of the Diamond Disk design (courtesy of Jon Lomberg)

REALITY BITES

These were the plans as of January 1997. Throughout 1994-97 the issue of how to write on diamond was a continuing technical puzzle. The final method adopted was etching by an oxygen and sand plasma. Oxygen etches all carbon-based materials, but optimizing this process demands much experience.

My doctorate was in solid state physics, so I was the natural person to deal with such issues. Quickly I learned how little I knew of technological advances in the last two decades; I was woefully out of date. We swiftly fell back on real experts. We got excellent technical support from Dieter Ast and Wendell Williams of Cornell University (who had been on the Pilot Project Marker Panel) and Paul Maker from the JPL Microdevices Lab. Lynden Erickson and his team at the National Research Council in Ottawa played the pivotal technical role, working with Professor N. Parikh of the University of North Carolina. Peter Taborek of UCI also contributed to early discussions.

Public exposure had already begun well before this work. One day in early February 1995, my telephone rang, announcing a leak. The British popular magazine *New Scientist* wanted to run a piece on the plans. Apparently they had sniffed out rumors from the European Space Agency. I guardedly confirmed what they already knew and corrected some errors. Immediately after hanging up I e-mailed Porco, Lomberg, and JPL. Porco seemed panicked, but I saw little harm. The local newspaper (the Orange County *Register*) also carried an extensive piece later in February 1995, by a reporter who sat in on our discussions at UCI. It provoked no follow-up journalism, and neither did the *New Scientist* piece.

Porco asked me to write *New Scientist* to correct their omission of her name, a request that seemed at odds with her anxieties about keeping a low profile. Still, I thought little of it. *New Scientist* published my letter, and a further note from Porco as well, establishing credit. Coincidentally, a paper I had coauthored on gravitational wormholes was getting enormous press coverage, appearing in over a hundred newspapers; the diamond marker drew surprisingly little interest. Wild ideas play well in the press, but as the launch date approached we expected the disk would get some mention.

I was working on the etching problem when Porco sent a stiff note in March 1996 thanking me for my efforts on the marker and dismissing me from further work. This was a puzzling shock. Porco was much exercised about the *New Scientist* story.

I reasoned that perhaps she was echoing NASA's extreme concern that nothing about a marker be made public before it was a done deal. The Pioneer plaque had provoked criticism, ranging from those perturbed by depicting nude humans, to feminists who disliked its showing a woman shorter than a man and in a different posture (less upright). In retrospect, the photograph eventually shot to go on the marker inevitably would have piqued some, since it showed a rather politically correct grouping of all races, with a woman as the centerpiece.

Porco's voiced concern over secrecy seemed strange at the time, since the *New Scientist* inquiry apparently had come from a European leak, possibly stemming from Porco's talk to the European Space Agency people. Prevailing NASA attitudes seemed to be that the marker was a vast secret, and as few people as possible should be consulted.

Getting the marker designed and made preoccupied Lomberg and me, not publicity. This proved naive, and these discordant notes resounded through all later events.

Since Lomberg and I had already laid out some major ideas, I was just as happy not to have to work through the principal remaining tasks: sharpening the concepts, detailed drawing of the disk etch pattern, and walking the diamond disks through the etching process. These first two fell to Lomberg, consuming many months of tedious labor. Porco handled most of the etching. Rather happily, I returned to theoretical physics.

Porco tried, but could get no money from NASA for any of this, so Lomberg and I had started looking for benefactors. By cutting me from the project, Porco freed me from this onerous duty as well, but Lomberg soldiered on, getting $50,000 from a prominent Japanese firm. From the beginning Porco had insisted that all funding had to flow through the University of Arizona, so she could control it.

Meanwhile, NASA was pondering our efforts. As with Voyager, our design team operated on its own, with minimal engagement of the busy engineers. However, Charles Kohlhase, manager for Science and Mission Design of the Cassini Program at JPL, decided that any marker or disk should carry no commercial insignia and issued a general directive stating so. When told of this, Porco commented to a colleague that she took her orders from NASA Headquarters. JPL is a NASA lab, and Porco was using her connections up at the top of the power pyramid. Though a mission scientist, she was independent of day-to-day JPL control.

At about this time we heard that the JPL Cassini group had begun to create their own marker. Previous missions all the way back to the Mars *Viking* lander, and perhaps even earlier, had carried the names of principal engineers, etched onto metal strips. Why not expand this idea and include the public?

With little time to spare, this Kilroy Was Here gesture could attract attention, public involvement, more hits at the Cassini web site. The Planetary Society joined in. Anyone who wanted their name to fly to Saturn had only to mail in a signed postcard. Signatures were cut out and scanned by the Planetary Society, then digitized and loaded onto a compact disk. After a national campaign roping in congressmen and canvassers, the grand result was 616,403 signatures on the carrier—named, in high bureaucratic style-deaf fashion, the Digitized Versatile Disk.

My congressman, Christopher Cox, sent all his constituents a letter promising to funnel their names through his office and onto the *Cassini* spacecraft: "Your name will live on in space long after your grandchildren, and theirs, and theirs." They obtained some celebrity signatures from *Star Trek* actors and congressmen, baby footprints and pet paw prints.

The European collaborators got wind of all this and started their own signature collection. They took the signature disk a step further and planned to sell a duplicate disk after the launch, reasoning that people who were to be immortalized on the interplanetary scale would, of course, want a copy. Like JPL, they set up a worldwide web site to send names and messages. They got such memorable phrases as "Hello green worms," "HELP," and "Don't cry because you cannot see the sun, because the tears will stop you seeing the stars."

The JPL team was uneasy about lack of screening of the ESA names and the Europeans' plans to sell their disk commercially. Therefore the European Space Agency attached their own disk to the *Huygens* lander, while JPL names fly on the *Orbiter*.

All this activity to collect a meaningless string of names and salutations emulates the portion of the Voyager Record of least value, the list of congressional committee members that NASA forced the Voyager team to include. The compact disk surely will not survive for more than a century or so, nor could it be easily read in any distant future. Even very clever humans or aliens could not figure out the encoding software from first principles, and even if so, they would get only a list of indecipherable, disorganized names, and a few cryptic, disconnected messages in this sea of words.

One could imagine a far future discoverer wondering what to think of a species that created a message without attempting to make it "comprehensible, self-extracting, anticoded, triply redundant, and graduated in content," as Lomberg summed up the Voyager and diamond disk approach. As a projection of pure vanity, it resembles the International Star Registry, which sells people certificates stating that stars have been named for them. Such meaningless exercises in ego tell more about our species than we might like revealed.

This Kilroy disk emerged only after the diamond marker idea became known. It had every sign of a hastily designed public relations stunt. Including long lists of names is a cliché of time capsules. Apparently, the largest collection was the twenty-two million assembled to be buried at the order of President Ford for the bicentennial celebration in 1976 (and then stolen, a classic irony). As Kohlhase put it to me, knowing that I looked askance at the signature disk, he and others devoted eighteen months of hard work to produce a "heart-based signature disk," in contrast to the "mind-based diamond."

Of course, both gestures spring from a common impulse: to give people a sense of connection with something larger than themselves. To value the "heart" is to rank the expressive quality of deep time messages over their communicating ability.

My trouble with all such name-gathering was that the end result more nearly resembled the graffiti that disfigure many ancient monuments. After all, the scribblers upon the Parthenon no doubt felt some burst of elation, too, but the end result besmirched the work that is the point of it all.

Lomberg regretted that the signature disk would get commingled in the public mind with the actual message marker, vastly increasing the ratio of noise to signal, as engineers put it. Indeed, the Planetary Society has now made this a feature of their membership drives; in 1997 they attached a microchip to the Stardust mission to rendezvous with a comet. "And you'll be a part of it all," an advertisement promised.

We can expect that such masses of names will become a standard fixture of a publicity-conscious space program. The Mars Polar Orbiter of 1999 will carry a signature disk, instead of the Visions of Mars disk crafted earlier. This decision came from a NASA lawyer's worries over sending any copyrighted material, despite permissions already obtained.

Shouting at the stars may become commonplace. In 1998 the Sci-Fi Channel tried to arrange transmission of signature messages by radio beamed skyward. An entrepreneur tried to sell space on metal plates to be launched to the stars. No one seems to have even thought about how utterly distinct life-forms could say something understandable to each other. Also in 1998 a French artist announced plans to launch a satellite on a long orbit to return in fifty thousand years, bearing art and messages from our era. This "archeological bird" will feature moving solar panels that will make it look like it is flying. With shape-remembering metal, it could carry sculpture, too. More adventurously, a firm advertised in 1998 plans to launch human hairs into interstellar space, at fifty .dollars a go. They anticipated getting several million customers who could send a hair sample (their DNA) and some small message as well. Perhaps bald men would feel left out, though there is nothing in the advertising circular that excludes pubic hair. A similar offer envisions broadcasting people's names into interstellar space by microwaves, a garbage SETI message perhaps to be beamed to the Andromeda galaxy. The hunger for Kilroy gestures seems unbounded.

So matters rushed on, as the launch deadline neared: October 1997.

COMEDY AND TRAGEDY

In early January 1997, Carolyn Porco sent NASA an agreement to sign which gave her sole control of the contents and public presentation of the diamond medallion. In February she e-mailed Jon Lomberg:

> As this project draws to a close, I don't think there is *any* question which of the two of us deserves the most credit . . . we are not "equal collaborators" on this Project. The major credit goes to me, and I would expect that anything you write on this subject honors my role. I intend to modify . . . anything that gets written . . . accordingly.

This was the last straw for Lomberg, who had originated the idea, designed most of the major elements, and had agreed to co-develop the disk with Porco. He had gone along with Porco's methods, seeing her as the NASA insider who would ensure that the diamond marker got to fly. As I soon thereafter wrote to JPL, "Lomberg probably knows more about the general issue of long-term messages than anyone, anywhere. Every major idea in the message either came from him or was significantly affected by his ideas and considerable work."

For some months Porco had communicated little to him of her work, keeping vital facts to herself. Lomberg saw her January move as an attempt to claim sole credit.

Nothing in the Pioneer or Voyager team experiences had prepared Lomberg for this. Charles Kohlhase at JPL knew little of Lomberg's work because Porco had seldom mentioned her collaborators. With this blowup, letters began coming in supporting Lomberg. At JPL, Lomberg found support from the project scientist Dennis Matson, Kohlhase, and one of the major scientists, Tobias Owen of the University of Hawaii. To the project manager, Richard Spehalski, they described Porco's behavior as "very troubling." When Spehalski booted the problem to NASA Headquarters, they tossed it back. Porco lobbied Spehalski, who removed himself from the process and returned the issue to NASA in Washington, D.C. By this time Porco had gone to the head of NASA, Daniel Goldin, to get his approval of the beach photo, billing herself as the "director" of the photo project as well.

Complications multiplied. Before, Porco had spoken of "large downstream revenues from merchandising" of the diamond motifs, perhaps to encourage DeBeers, which had duly delivered the disks for inscribing. Lomberg had warned her that based on his experience with the Voyager Record and the Mars compact disk, there was "a smidgen of glory but no money."

Now money raised another of its Gorgon heads. In the winter of 1997 a national political scandal had emerged; money had been funneled from China into the Democratic presidential campaign, a felony. Apparently, some at NASA worried if there was a parallel in the Japanese contribution to the Cassini message? Ever-political, NASA Headquarters became concerned about the Japanese involvement, though of course the firm came in because NASA refused to spend any of its own money. As one high official termed it, "we're troubled about fifty thousand dollars of Japanese money leveraging them into a three billion dollar U.S. mission." How much public exposure would the Japanese get? Would it look fishy to the press?

All this stirred in the bureaucratic pot. As the launch deadline neared, the soup heated up.

Porco would not budge on her demands, nor would she show her revised designs for the marker to anyone. Lomberg, as the principal designer, felt that he could not permit the message to be committed to the etching process until he had seen its final form. Given Porco's priority claims, he needed some guarantee from NASA that his three years of work would get proper credit. He decided to stand on the position that NASA "had to honor collaborative agreements made in its name by a scientist from an official project, acting for the project."

Porco would not give written guarantees of credit to the several collaborators, nor let Lomberg see her final design. Indeed, no one had seen it as January wore on. Porco claimed to have changed several components of the design, but did not display them.

"I have been leading all aspects of this project," she said repeatedly in e-mail to JPL figures, seeming to feel that leading in the bureaucratic sense implied playing the primary creative role. Indeed, this is common in large organizations, which routinely confuse administration with innovation.

Viewing this from afar, I felt doubly glad that she had tossed me off the project. Infighting, a NASA necessity, was not my style. NASA's final position, as expressed by Wesley Huntress Jr. was "You two must work this out for yourselves."

But animosity between Porco and Lomberg had long passed any hope of agreement, and NASA took no steps to make minimal assurances to Lomberg. They were barely communicating civilly. Throughout, NASA seemed more worried about the Japanese connection than with bickering among the creators.

On March 25 came the verdict from Huntress, by e-mail: "The Office of Space Science has decided not to fly the diamond-chip medallion proposed by Drs. Porco and Lomberg on the *Cassini* orbiter due to the lack of a clear agreement between the proposers on its implementation at this late stage in Cassini processing."

What lay behind NASA's curious dithering and then collapse? "Siding with the institutional insider versus the maverick outsider?" speculated one observer. "Taking the word of a scientist over the word of a mere artist?" Heads shook.

Surely part of Lomberg's resolve to dig in came from his conviction that eternal messages on behalf of our species should be

created by a process reflecting the values espoused in the message—
sharing, cooperating in amicable and peaceful ways, in contrast
with the promotion of a single figure. The contrast between this
incident and the brotherhood of the beach photo was too glaring.

Bruce Murray, a former head of JPL and now a professor at
Caltech, speculated openly that it was no coincidence that Porco's
sudden seizure of the message and refusal to divulge its contents
came so late. Carl Sagan, an old Lomberg collaborator, had died
only weeks before her move against Lomberg, and while alive
would "surely have intervened decisively." As a springboard to the
national stage, the diamond would be eye-catching. "In the end,
she appeared to want this more than she wished the marker to fly
at all."

Porco's high NASA connections had seemed an asset when we
were plotting how to get the marker on the spacecraft; now they
were clearly a liability. She had no fears of getting bounced off the
mission, Murray said, for as a former principal investigator on Voyager told me, "It's virtually impossible to get a NASA scientific
appointee fired. Despite major efforts and documented proof, I was
unable to ditch two incompetents on my team, and they were mere
underlings."

As Carl Sagan once remarked, "Ad astra per bureaucracia."

The news spread gloom. Over two years of work was wasted,
just weeks before the disks were to be inscribed. As an observer
had remarked when the Mars '96 craft went into the Pacific Ocean,
"In planetary exploration you have to enjoy the process without
fixating on the result."

Although NASA killed the marker idea, the pedestal for it had
already been installed on the orbiter spacecraft. It was a black rectangle in the proportions of the Golden Section. Further, it was to
have a round, golden case for the diamond and a copy of one side
of the diamond engraved on the case.

Porco attempted in the days after NASA's decision to write her
own, new message and inscribe it on the diamond coins. Since she
was principal investigator of the pedestal, the logic went, she had
a right to put anything on it she liked. By this time opposition to
her had mounted. Given the controversy that had already ensued,
JPL declined her efforts.

Porco also tried to stop any description of the marker project
from being published, writing to me to "forbid" it. Further, her

university returned nothing to the Japanese firm that had granted $50,000. Nor did DeBeers recoup the roughly $60,000 it had expended to make the disks.

The dizzying fall of so carefully conceived a project struck me as resembling the sudden collapse of a monument. I recalled the dictator Mausolus, whose tomb had become ornate with the best work of others, and ranked as one of the ancient Seven Wonders of the World. Overgrown with statuary, adornment, and fine sculpture, it was top-heavy and vulnerable to seismic shifts. Built on the vanity of a tyrant, in the end it was brought down in an earthquake by its own indulgence. A more wisely made monument would have better survived the tremors that go with its territory.

Louis Narens observed that Porco acted in a nonscientific manner, communicating less and less, conferring with fewer colleagues, and eventually refusing to show the final design to anyone at all. Instead of opening discussion, she closed it. This incident, he wrote, should provoke NASA to form an oversight committee to ensure that future deep space messages do represent the best thinking of the scientific and humanist community, and cannot turn into debacles.

This seemed to me, viewing the struggle from afar, the major lesson of the mess. JPL's Dennis Matson called the incident "a total management failure," and proposed that any plaques should be an integral part of missions from the beginning. Certainly, in view of the public reaction to the signature disk, NASA should form a coherent policy about designing and attaching any markers to them.

At the moment, as Louis Friedman of the Planetary Society remarked, their method is "ad hoc and secretly, as some type of afterthought. The subject doesn't have much priority in NASA, and therefore they try to have their cake and eat it too, by allowing it (which is laudable) but then by not dealing with it. . . ."

How should such messages be planned? I doubt that big official committees would display much ingenuity or speed. A free-for-all competition might be both fun and a better way to enlist the public's participation than through ever-larger signature lists. The ad hoc method sailed gracefully through narrow straits of money and time in the Pioneer and Voyager voyages, but with Cassini ran aground on the shoals of personality.

Who would judge such a competition? Perhaps NASA should keep a small, standing committee to either design messages themselves or to commission them, after entertaining suggestions from the large, clearly interested public. Not merely an elite specialist club, its members should have a broad understanding of the scientific and cultural elements that blend and inform each other in the construction of deep time messages. The challenge would be to include those understanding the physical and social sciences, creative arts, cryptography and language theory, religion and philosophy, yet keep the membership manageably small. Then some oversight committee could pass on the final design, or even stage a competition between several such subgroups.

Tobias Owen, another Cassini mission scientist, remarked that "the *absence* of a marker on *Cassini* provides a tiny, tiny hint of the dark side of the human psyche: That we don't get along with each other nearly as well as we would like (our cosmic neighbors) to believe . . . We are trying (as usual!) to tell ourselves how wonderful we are, hiding our blemishes as best we can. A message to the future is a perfect way to do this, as we can choose exactly what we want to put into it. But when we can't manage to send it . . . that is a dark message indeed!"

Cassini lifted off in late 1997, beginning its seven-year journey. Most of the media paid attention to the antics of a few dozen protesters who demonstrated against the use of seventy-two pounds of plutonium to power the spacecraft. The fact that such power sources have been routine for decades, and that three had already fallen back to Earth with no discernible harm, made no difference. The media needed a focus, and they like to present news as conflict, no matter how ludicrous an argument might be.

And after all, there was no message flying on *Cassini* to deflect their attention. Only then was the diamond disk missed at the public relations level; carrying along signatures provides little purchase for the news media.

Other missions will fly to other worlds, and NASA's increasing media awareness probably means there will be more messages attached. Well done, they may become long-term messages to our distant descendants, preserved in the vault of vacuum. Some shrewd judgment could shape NASA's future policies so that the messages carried convey something this civilization truly wishes heard across the great silences of uncountable epochs.

That is perhaps the best lesson learned in all this: we should think well and hard before we act again.

As for Cassini, in the end the conspicuous, empty mounting plate was an embarrassment. In a fine point of irony, the pedestal was then coopted by JPL's Digitized Versatile Disk, to carry its 616,403 signatures: Kilroy triumphant. All that remains of the diamond marker and the ideas behind it is the pedestal, now on its long arc to Saturn.

PART THREE

THE LIBRARY
OF LIFE

Epithet after epithet was found too weak to convey to those who
have not visited the intertropical regions, the sensation of delight
which the mind experiences . . . The land is one great wild, untidy
luxuriant hothouse, made by nature for herself.
— CHARLES DARWIN
THE VOYAGE OF THE BEAGLE

In the broad sweep of millennia, our era is unique. The press of human numbers, plus our wasteful predation, now threaten to doom a sizable fraction of all the world's species within the next century.

Though, like most people, I thought of myself as an environmentalist, my own slowly dawning realization of these blunt probabilities came only in the early 1990s, as I became acquainted with the work of E. O. Wilson and others. Many details of this subject are hotly contested, but I accept the plausibility of a genuine crisis approaching, a broad calamity from which the bulk of humanity will not escape unharmed. Our rise in population is startling when seen on the scale of truly deep time.

This intersects our interest in deep time messages, for there is no more firm message than extinction. When we deprive the future of a species, we send forward a blankness in the weave of life. Nature adjusts to that absence, threads fill in the weave, and generations to come dwell in a truncated world made poorer in ways difficult to predict.

Asked what evolution taught us, J.B.S. Haldane replied, "That God has an inordinate fondness for beetles." Few know or care about the fate of a specific type of beetle, for there are so many; but we know little or nothing of the impact of even a single such loss. Far more portends for us, emotionally, from the loss of large,

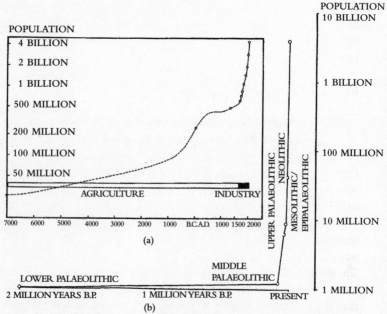

Figure 3.1 (a) Growth over the past nine thousand years, the span of civilization and history. (b) Growth of the human population over the past two million years. (courtesy Warren Hern, *Population and Environment,* March 1990)

charismatic beasts; I regret that I shall never see a mammoth or a dodo, which our kind quite probably drove into oblivion.

The coming wave of extinctions, then, may be our most lasting, important heritage, far surpassing our gilded monuments or abstract plaques on spacecraft, for the vast loss will be felt throughout humanity in a world truncated and diminished, perhaps irretrievably.

This message will permeate billions of lives, literally at the ground level. It will be an unconscious message, unnoticed, for most species shall die by being crowded out while we are about our busy business; but it shall be no less effective.

Before reviewing the case for conservation, and thus the need to send a radically different message across centuries, I shall go directly to the proposal I made in 1992 to enlist the strategies of deep time messages in this cause.

This was and is a radical idea: to convey a new kind of message, intensely information-dense, a signal of desperation. The target lies at least a century away, perhaps much longer: nothing less than a future generation that needs the information lost in our coming dieback of many species, and can harvest our salvaged samples with technology we cannot foresee. My eventual hope is that the method will become standard, even in less endangered times.

If we do not save some record of the species now vanishing, their loss will reverberate down through millennia.

SALVAGE BY SAMPLING

Our situation resembles that of a browser in the ancient library at Alexandria, that great repository of classical culture ultimately put to the torch by religious fanatics.

Our browser suddenly notes that the trove he had begun inspecting has caught fire. Already a wing has burned, and the mob outside seems certain to block any firefighting crews. What to do? There is no time to patrol the aisles, discerningly plucking forth a treatise of Aristotle or deciding whether to leave behind Alexander the Great's laundry list. Instead, a better strategy is to run through the remaining library, tossing texts into a basket at random, sampling each section to give broad coverage. Perhaps it would be wise to take smaller texts, in order to carry more, and then flee into an unknown future.

While other efforts to contain and control our accelerating biodiversity disaster are admirable and should be strengthened, it may well be time to consider a similarly desperate method of biological salvage.

I proposed that the biological community ponder a systematic sampling of threatened natural habitats, with long-term storage by freezing. This would more nearly resemble an emergency salvage operation than an inventory, for there would be minimal attention paid to studying the sample. The total sample mass might be reduced by judiciously trimming oft-repeated species of the prolific ants and beetles. (Duplication in sampling makes for good statistics, though, helps comparative anatomy, and may aid those for whom more successful species are more interesting.)

The essential aim is to save what we can for future generations, relying on their presumably better biological technology to extract the maximum benefit. This is a different sort of deep time message, neither High Church nor Kilroy Was Here—unless by the Kilroy autograph we mean the mute signature of a species, its genetic record.

None of this proposal would involve new technology. Sampling of tropical trees by insecticidal fogs and active searching of the canopy is common. Teams trained to simply collect, without analyzing, require minimal labor by research biologists.

Freezing at the site can be done with ordinary ice or dry ice; liquid nitrogen suspension can occur only at the long-term repository. Extensive work by taxonomists enters only when samples are studied and classified. Here lies our current bottleneck. There are far too few taxonomists to tally the world's species within our generation, let alone analyze them.

We sidestep this problem if our primary aim is to pass on to later generations the genetic essentials of our immense, threatened biodiversity. Extensive taxonomic expertise would defeat the project. Detailing the many variations within a population probably will not be worth the trouble. Even information about the existence of a species is useful, for without a sample, in the future one cannot be sure whether a given variation ever existed, or simply became extinct without being observed.

It seems likely that captive breeding programs, parks, microhabitats, and zoos can preserve only a tiny fraction of the threatened species. Here I shall use the term "preservation" to mean keeping alive representatives of at least each genus—in situ, in vivo (living, on the spot) protection in reserves—and later argue that this is essential to eventually studying and potentially resurrecting frozen species.

To save the biosphere's genome heritage demands going beyond existing piecemeal strategies of seed banks, of germ plasm and tissue culture collection, and cryopreservation of gametes, zygotes, and embryos; these programs mostly concentrate on saving traditional domesticated varieties. Our goal would be a broad, admittedly incomplete sample of all threatened species, suspended in liquid nitrogen for the indefinite future.

I wrote a paper proposing this, and in 1992 submitted it to several scientific journals. *Science* and *Nature* rejected it, but with the help of Frederick Reines of my own physics department (who was soon to win the Nobel prize), I sent it to the *Proceedings of the National Academy of Sciences*. After favorable referee reports, they published it essentially unchanged, and ignited a considerable controversy. To see why, I must outline the larger arguments.

THE CASE FOR A CRISIS SOLUTION

Though botany and biology are millennia old, we have only begun the elementary taxonomic description of the world biota. While we have given about 1.4 million species scientific names, estimates of the total number of species range up to roughly thirty million or higher. Our recent discovery that an entire third kingdom of life exists beneath us, the ancient Archea, means that we are even further from understanding the living wealth surrounding us. (The other two kingdoms are bacteria and all those carrying nuclei in their cells, including us.) We may very well not know the species diversity of the world flora and fauna to the nearest order of magnitude.

Time is running out in which we can even catalog our living wealth. The overall natural extinction rate from fossil evidence is about one species lost per decade. A very conservative estimate of the current extinction rate gives roughly five thousand species lost annually, with some values far higher.

Worse, this rate is rising, accelerating us toward a calamity with few parallels in planetary history. (Asteroid impact probably did a quicker job.) The best known cause of present day species extinction is the cutting of tropical forests, which have lost about 55 percent of their original cover and are shrinking at the rate of 1.8 percent per year. Worldwide, one in every eight plant species is threatened with extinction, the first broad survey showed in 1998. With a ratio of one-to-three, the U.S. is the world's worst case.

Growth rates seem doomed to increase, since their ultimate cause is human activity, and human numbers and expectations grow apace. To improve the lot of a swelling human tropical population would require at least a fivefold increase in economic activity there in the next century, a crushing load on the already strained biosphere.

Other biological zones such as coral reefs and oceanic islands also dwindle at alarming rates. Losses are most severe in precisely the species-rich tropical continents where our own numbers swell so alarmingly. As Paul Richards remarked in *The Tropical Rain Forest*, the region "has a fatal tendency to produce rhetorical exuberance in those who describe it," equaled now by an offsetting expressive gloom at future prospects.

Everywhere there are calls for a halt to tropical deforestation, but most knowledgeable voices seem tinged with despair. Paul Ehrlich and E. O. Wilson suggest we could lose a quarter of *all* species in half a century, with incalculable effects on our biosphere. We now coopt about forty percent of plant growth worldwide, favoring monocultural crops, which must greatly reduce genetic diversity. Given the blunt economic and cultural forces at work, even slowing the rate of destruction seems doubtful in the immediate future.

Nor is favored North America all that well off. Almost a third of all plant species in the U.S. appear to be at risk. Though California has more plant species than any other state, it ranks behind only Hawaii in total number at risk, with Florida coming third. California has many diverse habitats, giving rise to many narrowly restricted species, vulnerable if they live at the margin of the state's spectacular urban and agricultural growth.

Increasingly we North Americans follow the lead of the truly besieged regions, such as India and Africa: parks for animals, the rest for people.

But few natural parks can sustain viable populations of large animals, especially carnivores. Whole continents can prove too small; lions would live in Europe's southern climes, but they could not move northward after the last ice age because we got there ahead of them. Even prestige animals like tigers probably cannot survive the next century without extensive backup populations in zoos. Parks nestled into a continent cannot represent its range of life, especially the larger mammals, because the included populations are too small to sustain themselves genetically. Breeding members produce few offspring, or stillbirths.

Already some park bird species are mainly old birds, like the cockatoos of Australia. Fully half of Asia's elephants are old and will probably have no descendants. Similar situations strongly suggest that we shall see quite sudden diebacks in the early twenty-first century.

"Edge effects"—encroachment of weeds, pollutants, cattle, feral cats, and enterprising people—threaten even large parks. What biologists cozily call "species relaxation" occurs: the slow, steady loss of numbers, despite apparently good living conditions in the contradictorily termed "controlled wild." This has led some biologists to concentrate heavily on fully integrated studies of landscape and habitat, while others try to save particular species. Both approaches have virtues, but measures such as creating corridors between habitats, to allow larger populations to circulate and breed, cannot offset edge effects, and in fact are particularly vulnerable to them.

A currently popular standard calls for protection of about ten percent of the total land area in each nation or ecosystem. Using the relation between species number and area, this implies that about half of all species would be vulnerable to extinction.

Throughout the 1990s the rate of tropical deforestation persisted at about 0.8 percent per year, while the rate of nature reserve creation fell. Most governments are spending on "sustainable development" experiments, rather than strengthening the global system of protected areas.

ZOOS AND FREEZERS

Zoos become crucial at this point. Increasingly, biologists now turn to *ex situ* (away from the site) conservation strategies to save the world's biodiversity. For many years zoos merely entertained by bringing the untamed wild to a gawking public that watched from safety. Now they often protect the wild from untamed civilization. Zoos and other conservation organizations are all too often the last refuge for rare and endangered plants and animals. Their occasional success is perhaps epitomized by the recent comeback of the California condor, now not quite so close to the brink of extinction.

Captive tigers already outnumber those in the wild, and many other species can see this on their horizon. Zoos produce animal populations that differ from the wild state in both behavior and genetics, though "arranged marriages" carried out with sophisticated genetic inventories of the couple can offset some of this. Some purists sniff at such methods, but I fear that, as biologist Colin Tudge remarks, "The next few centuries at least will be desperate times for all of us; and desperate measures are called for. . . ."

Zoos are unhappy solutions, for many; some make us ashamed of our own species. In Rome the tiny, dirty cages bear no identification at all for many of the visibly miserable, bored residents. In many zoos, strained for space, large animals (even bears) cower in small spaces. The otherwise excellent San Diego Zoo keeps its polar bears in a gunite enclosure with a stagnant, unappetizing pool. The latest rage for preserving the rain forests has led to exhibits directed mostly to making the patrons think they are in a rain forest, but not giving the animals such illusions, since many are caged. Only aviaries seem to avoid this trap. In many zoos I have seen animals like wild fishing cats in tiny cages that leave no room for these shy, elusive creatures to escape prying eyes.

After peering into countless sullen, distant, or anxiety-ridden eyes, one is unsurprised to learn that the drug firm Eli Lilly has donated free Prozac to a zoo as an act of mercy.

The Toledo Zoo made an estimated $60 million in 1988 from five months' residence of two pandas. With fewer than a thousand left in the world, this level of money means they will be treated as cash cows first and as conservation subjects second. (Indeed, a brand of utilitarian argument goes, consider how many more species one can save with the panda income. Regrettably, we must risk one species to save dozens more.)

Of course, we cannot ask the animals how they feel about all this (though some inquiries with chimpanzees in our species communications programs might be revealing). Animals do live about twice as long in zoos as in the wild. Quality of life is difficult to discuss; the inability of many species to reproduce in captivity suggests a persistent anxiety the animals may not manifest to the casual eye.

Zoos are uncomfortably aware of these points; some have started relabeling themselves "conservation areas." Others put their few dollars into captive breeding and even re-release programs. These are important, for worldwide zoos house only about half a million individual mammals, birds, reptiles, and amphibians. In the long run, as concerns us here, large populations of fewer than a thousand species can be cared for in zoos. Saving the last few animals of a species near extinction is, of course, crucial, and here they do very well. But animals have gone extinct in captivity, too.

How about preserving plants through their seeds? Here prudent agriculture authorities have had major impact. Worldwide there are more than seven hundred documented seed collections (some chilled, some not) holding an estimated 2.5 million entries, including many exotics. The U.S. alone spends over $20 million a year on germ plasm acquisition and preservation. But collections themselves are threatened worldwide by dwindling budgets. A unique all-taxa survey of Costa Rica recently collapsed.

Even among successful collection efforts there are major differences of philosophy about how to sample in nature: whether to select for genetic composition rather than simply for appearances. But sampling unknown populations means that one cannot know the genetic features in advance. Most seed banks concentrate on gathering and saving useful crop plants, particularly grains. While useful, this leads toward a monoculture inventory, with much data on grains and little about the larger world. And seeds die. Even in cold, dry storage, few are viable after a decade.

BEYOND ENVIRONMENTALISM

Environmentalists embrace the concept of sustainable consumption through using renewable resources, a fine idea in general. Grudgingly we must realize that this has little to do with species preservation. At best the sustainable level of human-driven extinction is very small, and indeed, unknowable. We have been altering the biosphere through extinction ever since we homo sapiens came on the scene roughly fifty thousand years ago.

How pleasant it would be to argue for protection of all species out of respect for their beauty, their role in a complex system we do not fathom yet (and may never), or for the simple spiritual reason that they are the work of some Creator. But seldom has society supported these motives for very long, or at all. Most preservation cases provide an economic payoff as well as the aesthetic, ethical, and spiritual ones.

Leading off with economic arguments is perfectly fine, since they are most likely to win acceptance, so long as we do not see economics as forever primary. Our cultures have no sure grasp of how to balance moral and aesthetic sensitivities against economic prosperity. So long as the poor proliferate, many will feel they must come first. But the poor are always with us, while many species won't be.

The great social task confronting us is the uplifting of the bulk of humanity to a decent living standard, in the face of surging populations. Here cost-benefit frameworks may work well in the short run, but not in the long one. Clearing land for crops and thus losing species helps those present and hurts humanity in later generations.

How to impress our culture with this long distance, deep time prospect? Generally, one plays to eye-catching issues.

One aspect of widely supported environmentalism is the saving of big mammals. Everybody supports keeping the cheetah and elephant and gorillas around (not that these outcomes are guaranteed). Hard-liner biologists sometimes sneer at this impulse to save "charismatic vertebrates."

There is a sound argument to support our intuitive sense that large beasts matter. Big animals are far more vulnerable than they seem. First, there are fewer of them, because they need more area to support their diet. Second, they have few offspring, investing in quality rather than quantity. Finally, they are conspicuous, tempting targets for hunters, so "charismatic vertebrates" are especially exposed in the shooting gallery of the wild, a constant fitness sweepstakes.

There is some evidence that the hunters of the Upper Paleolithic were not just obviously good hunters, but game managers as well. They used fire to drive herds, as do the Australian aborigines. Over many millennia they seem to have sustained useful levels of the animals they hunted. When human numbers swelled, though, or people moved into fresh territory, matters often got out of hand; the preservationist agenda failed before expansionist tendencies.

Throughout the Cenozoic, most mammal species have lasted roughly a million years before going extinct or evolving into another species. In this sense we humans, a mere fifty thousand years old, are near the beginning of our run. The next century will show whether we can live in equilibrium with a large fraction of our fellow life-forms.

The price for failing may well be our quick extinction. As Colin Tudge has remarked, "Nature's rules are universal, and nature has no taste. We may find our own collapse unthinkable simply because it is so horrible to contemplate, but repugnance provides no protection at all, any more than incredulity protects against flood or the encroachment of ice."

Agriculture, our greatest invention, has allowed our numbers to increase a thousandfold in ten thousand years. How much can the Earth support sustainably? Conservation biologists' estimates range from 0.3 to two billion. Today, 300 million is barely more than the current U.S. population, but it was all of us at the time of Christ.

Why care if we crowd out other creatures? Self-interest alone argues that other species' genes are invaluable resources, as both breeders, farmers, and biological engineers know. We cannot merely protect those genes we know to be useful, either, for that catalog steadily expands as we learn.

Aesthetic arguments—save the tiger, he's so beautiful—are vaguely materialistic, for they assume that we suffer animals to live if they please us. Ethical issues then arise: as the most powerful species, we are the rightful stewards—indeed, the only possible ones.

There is a muted form of this, advanced by the World Conservation Union: minimize our regret. Since we do not know how a world stripped of species will work or look, we should prefer a rich future to an impoverished one. Once extinct, elephants might well never reevolve. This is also an argument for some way of preserving some "essence of elephant," for to do so helps minimize regrets.

So we must save wilderness—a term that in policy circles invokes everything from random patches of green up to parks the size of small countries, like the U.S. Yellowstone and South Africa's Kruger. But all reserves are the creatures of governments, quixotic stewards at best. Ever since Tom Paine's debate with Edmund Burke over the French Revolution, western societies have generally felt that no generation has the right to bind all future ones in perpetuity.

Nature itself cannot conceivably correct for our excesses. Surely our headlong rush to an immense dieback in the next century allows no time for evolution to adapt species to us. Shrinking habitats sometimes force evolution of dwarf species, for example, but that takes perhaps thousands of generations or more.

Altering behavior is quicker and far more likely. The American bald eagle has returned to many areas because it has gotten used to us. To some extent these changing habits may stem from losing the habit of fleeing all humans. We have taught animals for many millennia that we kill from afar. Now many of us instead stare in awe at a wondrous eagle. The birds no longer need flee automatically, so that trait is ebbing.

Amid this "human bloom" we must keep our wits about us.

Particularly, conservation is not a politically polarized left-right issue. Barry Goldwater was a conservationist all his life.

Further, we should recognize that proponents of indefinite expansion often use the poor as a shield. A new "isolated" resort or
highway does give jobs to laboring people. Economics does not
grant that other species exist; all branches, from free marketeers to
Marxists, are resolutely anthropocentric. "The best things in life
are free" can mean that, being free, they are not weighed in the
balance, and hence get exploited.

Intuitively, we prefer direct, in situ preservation of endangered
species. The broad public, and probably most biologists, think in
these terms, with some appreciation for the conservation aspects of
zoos and aquaria. But two liabilities of in situ measures become
apparent as our global crisis worsens: cost and reliability.

No one proposes simply buying up all the tropical forests. The
huge cost is the least difficulty; what will deflect the vast social
pressure of a burgeoning human population? How long could even
the legal owners hang on to vast landscapes? Property rights bend
in the prevailing winds of social change.

The hardworking farmer with a family and a chain saw is an
often unknowing enemy of biodiversity. His modest economic ambition makes many in situ schemes too expensive, economically
and politically. This pressure leads to long-term uncertainty in
holding the line of even the best local programs, as in Costa Rica.

Furthermore, the price of in situ is eternal vigilance. Inevitable
defeats will visit upon us the loss of uncounted species—beetles and
bushes, worms and willows—which we protected for only a while,
often without even knowing of their existence. Is there a way to
hedge our bet?

This dire moment demands radical thinking. In the spirit of a
thought experiment, I sought in my 1992 proposal to link the in
situ preservation community, which emphasizes protected wild areas, and the *ex situ* conservationists, such as zoos, botanical gardens,
seed banks, etc. For in situ measures there are economic, environmental, and aesthetic arguments. To preserve the genome of many
species, however, *ex situ* methods may suffice. Considering this
possibility serves to separate the kinds of arguments we make for
conservation methods, including concepts of our moral debt to
posterity. In the spirit of sharpening debate by considering plausible
scenarios, we can test our ideas.

WHY FREEZE?

Banking cells by drying them with silica gels is useful for short times, but when kept at room temperatures, thermal damage to DNA will accumulate over the decades.

Much more than data about existence can be carried forward by simply preserving a wide sample of life-forms, held at low temperatures. We know that seeds can germinate after lengthy freezing, and that microbes can sustain cryogenic temperatures. Single cells such as sperm and ova survive liquid nitrogen preservation and function after warming. Generally, organs with large surface/volume ratios preserve well, such as skin and intestines.

Of course, more complex systems suffer great freezing damage. Research into minimizing this is making steady progress. Several kinds of damage occur, and little is known about methods of reversing such injury. Biochemical and biophysical freezing injury arises from shrinking cell volume as freezing proceeds. Plants ooze lipids as their cells contract during freezing. When thawed, those lipids do not spontaneously return to the expanding membranes. Restoring the normal living volume proves impossible. There are ways to reverse this, but the tissues end up depleted of membrane proteins. Major fracturing of cells and their extensions such as axons and dendrites, or capillaries and other elements, causes extensive damage at temperatures below the "glass point" where already frozen tissues harden further and shear begins. For some purposes, then, immersion in liquid nitrogen may be unacceptable.

The problem of recovering viable cells from frozen samples is complex, but even low survival rates of one cell in a million are good enough if the survivor cells can produce descendants. However, our minimum aim can be to simply retain DNA, the least we should expect from a sample. Of course, suspending whole creatures retains far more information.

For this, liquid nitrogen is suitable for long-term storage ($-196°$ C), especially since it is by far the cheapest method. At twenty-five cents per liter, liquid nitrogen is the lowest-priced commercial fluid, after water and crude oil. It allows suspension in large, easily tended vaults, simply by topping off the amount lost. Only a wholesale breakdown of industry can plausibly destroy the samples; no mere power failure will do. Redundant storage at different sites avoids even this.

Further, while neither liquid nitrogen nor freeze-drying damage DNA, freeze-drying does cause far more injury to structural and taxonomic characteristics. For the broad program envisioned here, which should also include small samples of ocean water with its teeming viruses and bacteria, plainly, liquid nitrogen is essential. This is also true for saving whole creatures, since we also gain their parasites, bacteria, and viruses, which are better preserved cryogenically.

A crucial point is that we need not rely on present technology for the retrieval. Progress in biological recovery can open unsuspected pathways. In a century we will undoubtedly know far more.

Recent advances underline this expectation. Techniques such as the polymerase chain reaction can amplify rare segments of DNA over a millionfold. Similar methods have enabled resourceful biologists to recover specific segments from such seemingly unlikely sources as a 120-year-old museum specimen, which yielded mitochondrial DNA of a quagga, an extinct beast that looks like a cross between a horse and a zebra. A five-thousand-year-old Egyptian mummy has yielded up its genetic secrets.

Amplifiable DNA in old bone is beginning to open study of surviving organic matter from the deep past. The current record for bringing the past alive in the genetic sense is DNA extracted from a magnolia leaf trapped in lake mud between seventeen and twenty million years old. This feat defied a prediction that DNA could not survive intact beyond about ten thousand years.

We should recognize that future biological technology will probably greatly surpass ours, perhaps exceeding even what we can plausibly imagine. Our attitude should resemble that of archeology, in which a portion of a site is deliberately not excavated, assuming that future archeologists will be able to learn more from it than we can.

Nonetheless, our choices of what to preserve cannot depend upon unknowable future capabilities. What rules suggest themselves from basic biology?

PRESERVE THE GENUS, FREEZE THE SPECIES

It seemed to me that to salvage biodiversity out of catastrophe, the best approach may be two-pronged:

(a) preserving alive some fraction of each ecosystem type ("biome"), its population intact at the genus level, and

(b) freezing as many species related to the preserved system as possible

A genus is the next step higher in classification, grouping similar species. It is a rough concept, only qualitatively defined; future generations of biologists may refine it, in ways difficult now to guess.

Freeman Dyson has suggested to me that a better notion is to define a *clan*, defined by the operation whether a sample of extinct creature A can be brought to life in living creature B: A → B. (Surely this is a technology dependent, and thus time-dependent, definition.) If for another creature C we have also B → C, then clearly if we have C alive, we can resurrect first B, then A. A clan is all those creatures A belonging to resurrection chains with the same root C. Clans may overlap, and future taxonomists will have to agree on a canonical list of maximal clans that cover all known creatures. One of our aims in a Library of Life would be to capture at least one root population for every maximal clan.

Here I shall use "genus" rather than "clan" to denote this transferability, because genus is a well-known concept. I am suggesting a concept for which we have no good term. Transferability has been studied in mammals, with much less done for other classes. For birds, there would be obvious troubles arising from the different nesting behavior, different egg sizes and hatching times, and the like—not all amenable to the standard incubator methods.

Carrying this research into reptiles, amphibians, and fish seems a large enterprise, suitable for work for the next century. The size of a clan among the invertebrates might be enormous, since they share ancient roots, though the social insects are a separate case.

At a minimum, this strategy would allow future biologists to extract DNA from frozen samples and study the exact genetic source of biodiversity. Genes of interest could be expressed in living examples of the same genus, by systematic replacement of elements of the genetic code with information from the frozen DNA. Obviously, the preserved genus is essential.

These techniques would open broad attacks on the problem of inbred species. Population bottlenecks—low numbers of individuals caused by isolated or fragmented habitats, or by a ravaged environment—can constrict the genetic diversity of an individual species. Reintroducing diverse traits from frozen tissue samples could help such a species blossom anew, increasing its resistance to disease and the random shocks of life.

Beyond this minimum—the DNA itself—future biologists will probably find great use for recovered cells in reexpressing a frozen genome. Using cells to resurrect mollusks, trees, insects, and other species is a cloudy, complicated issue; we know little about it.

For mammals, uterine walls and elements of the sexual reproductive apparatus should prove essential. After all, using present day sperm and embryos demands much collateral information. Here, using host mothers of the same genus is crucial. We should not rely upon our descendants to resurrect creatures straight from reading their DNA!

We must face a blunt fact: at reading genetic information, we are at best marginally literate, hoping that our children will be better readers, and wiser ones. Many biotechnological feats will probably emerge within a few decades—many ways, let us say, of reading and using the same genetic "texts." But no advanced "reader" and "editor" can work on texts we have lost.

Selectively reintroducing greater biodiversity in the future could gradually recover lost ecosystems. Individual species can be resurrected from very small numbers of survivors, as the nearly extinct California condor and black-footed ferret have proved. Frozen genes could also increase genetic diversity in such populations; some emergent programs do so now.

Fidelity in reproducing a genome may not be perfect, of course. Many practical problems arise such as placenta environment and chemistry, for example, complicating expression of a genotype. Perhaps animals birthed through another species of the same genus will incur some unavoidable variations. In any case, future generations may well wish to edit and shape genetically those species within an ecosystem as they repair it, for purposes we cannot anticipate.

Loss of nearly all of an ecosystem would require a huge regrowth program, for which the Library of Life would prove essential. Suppose, though, that we luckily manage to save a large fraction of a system, through conventional land preservation. Then the species library will provide a genetic "snapshot" of biodiversity at a given time and place, which evolutionary biologists can compare with the system as it has evolved much later, through many generations. This would be a new form of research tool.

Already a crash program to collect permanent cell lines, DNA, or both from vanishing human populations has excited attention. This program maintains cell lines by continuous culture, a costly method that invites random mutation. Such records may allow a deeper understanding of our own origins and predispositions, but banking frozen tissues of endangered species is the only way to ensure that any genetic disease diagnosed in the future in small, closed populations (the "founder effect") can be mapped and managed. In 1964, Paul Ehrlich suggested the creation of "artificial fossils" in such fashion. The "frozen zoo" of San Diego, begun with this in mind, has immersed more than three thousand mammal fibroblast cell cultures and several hundred tissue pieces in liquid nitrogen—over three hundred species, in all. Cryonic mouse embryo banking for genetic studies is now routine.

All *ex situ* programs have liabilities, just like in situ, arising from the brevity of human concerns. Programs from seed banks to systematically frozen samples have all failed because a career ends or funding dries up, freezer space gets scarce or zoos lose interest. However, the cost of frozen *ex situ* salvage is far less than an in situ strategy covering the same number of species.

In any case, a deliberate, full-scale *ex situ* attack can plausibly survive the erosions of time far better, for it is compact, inexpensive, and easily protected. Alas, none of these applies to preserving large land areas against the "human bloom."

The far larger prospect of eventually reading and using a Library of Life is difficult for us to imagine or anticipate, since we live in the early stages of a revolution in biological technology. Our situation may be likened to that of the Wright brothers if they had tried to envision a moon landing within three generations.

SPACE AND SPECIES

While it would be ideal to collect specimens from all parts of every habitat, economic constraints are profound.

Where should such sampling be concentrated? Clearly, areas with high diversity and deemed most endangered should be sampled first. Beyond that, species-rich locales such as riverbanks and ridge lines should draw preferential sampling. They are boundary communities that trace the interaction of species from different habitats. Narrow land necks are also important; Central America covers less than half a percent of the world's land area, yet holds seven percent of its species.

In the northern hemisphere, the Nature Conservancy uses a comprehensive system that describes forest types and composition, to establish priorities for preserving biodiversity. No such system exists for the tropics, as most forests are described by physiological characteristics (such as rainfall) rather than composition. Our knowledge grows slowly, and new remote sensing technologies are providing increasingly greater detail about vegetation formation, and even structures beneath the canopy. Satellite photography yields data about areas being heavily logged and burned; these might be given special consideration.

A globally systematic but locally random sampling might follow the criteria of biologist Vernon Heywood:

1. The area is species rich.
2. The area contains a large number of plant species restricted to the local area.
3. Sites contain important gene pools of plants valuable to humans or of potential value.
4. Sites include a diverse range of habitat types.
5. Sites contain a significant proportion of species adapted to special local conditions, such as peculiar soils.
6. Sites are threatened or are under imminent threat of development.

Several salient questions about even such broad criteria immediately occur: Should we limit ourselves only to sites that are species rich? Should we assume we know what the future will deem important?

Modern biology is young, only beginning to explore its deep genetic fundamentals. This suggests that we pursue strategies unencumbered by our narrow present concerns. Often, "species rich" does not include microbial densities. The concept of species is itself a bit slippery, certain to be refined and altered in the future. Further, current programs, such as sampling of tropical trees by insecticidal fogs and active searching of the canopy, may be most useful if they take everything, without laboring under species-specific goals.

Generally, rainfall and vegetation increase closer to the equator. Species numbers rise sharply upon entering the tropical belt, at a latitude of 20 to 25 degrees. Forests covering much of this region, between the tropics of Cancer and Capricorn, hold a wealth of unique diversity.

The World Wildlife Fund has identified "megadiversity" countries: Mexico, Colombia, Brazil, Zaire, Madagascar, and Indonesia. These countries hold a major portion of the world's biodiversity and are most at risk. The U.S. National Academy of Sciences Committee on Research Priorities in Tropical Biology identified eleven areas deserving of special attention because of their high diversity, large number of locally adapted species, and rate of forest conversion. These were: coastal forests of Ecuador; the "cocoa" region of Brazil; eastern and southern Brazilian Amazon; Cameroon; mountains of Tanzania; Madagascar; Sri Lanka; Borneo; Sulawesi; New Caledonia; and Hawaii. A large scale sampling program should first look to these.

Much attention focuses on the tropical rain forests, but in truth they are not the most endangered. That dubious honor belongs to the tropical dry forests. Dry forests once occupied more than 550,000 square kilometers from Panama to western Mexico. Today less than 2 percent remains intact, and only 0.09 percent has official conservation status. Dry forests endure a four-to-seven-month dry season, which makes them easy to clear by fire.

Biologist Daniel Janzen has argued that conservation strategies should not solely attempt to preserve the largest numbers of species possible. He argues that the creatures of the dry tropical forests are intellectually more interesting in their interactions, and thus more worthy of protection. As well, wet and dry forests share migrants and are ecologically interdependent. Obviously, we all bring to conservation our preferences. A virtue of a mass, random sampling is its impartiality.

Clearly, any large sampling effort should consider issues other than just obtaining maximum numbers of species. The effort should try to capture "biodiversity" in the broadest sense, lest we forget critical areas that might not be the hot topic of the day.

Desire for a complete catalog of species is widespread, and the Library of Life project would appear to be an excellent opportunity to begin classification of all the world's species. Only by attempting this on a small scale can we establish the methods for the larger problem.

But complete classification is both economically unfeasible and logistically impossible. E. O. Wilson, arguing for a complete identification of the world's species, calculated that 25,000 professional lifetimes would be needed to identify ten million species. If his estimate is correct, and each lifetime cost only a million dollars (a conservative figure), the total expense would be $25 billion! Clearly, taxonomy is far more expensive than frozen salvage. To be fair, Wilson argued that computerized identification methods could speed up the process, but at the same time, ten million species might be a gross underestimate.

Even if classification were cheap, other problems make it impossible. There are fewer than fifteen hundred professionals in the world with expertise in tropical species, and even this number is rapidly declining. Taxonomists are moving to better paid positions, and are not replaced when they retire or are laid off from museums and herbariums.

Quite probably, even a much larger team could not keep up with the tremendous volume of material obtained from an aggressive sampling project, such as the library proposal. John Terborgh illustrates the problem:

> Even when specimens are in hand for each of the 600 to 800 trees that typically occupy a hectare, the job is far from done. The specimens must be taken back to a world-class herbarium and then sorted, first to family and then to genus. The specimens are then packaged and mailed out, genus by genus, or family by family, to perhaps the only human being competent to identify them. Likely as not, the specialist is burdened with dozens of such requests, each of which must await its turn. It might be a year or two before the specimens come back with names sanctioned by the expert. Specimens lacking flowers or fruits often cannot be identified at all.

The problem is worse for insects. We do not know how many species of beetle there are even to the nearest order of magnitude! Quite likely, many species will appear hundreds or thousands of times in the sampled collection. This is a virtue, in that we get a rough measure of the area and density distributions of species, but bare taxonomy of them would be pointless. Repeatedly classifying the same species pushes the cost of identification painfully high. Keying out hundreds of millions (perhaps more) of organisms could take centuries.

Then, too, the usefulness of such information is itself in doubt. Our current species classifications may not weather the winds of change, as genetic analysis redefines our terms.

The point of sampling a vast array that one does not know in detail is to capture the raw animals and plants whole—not just DNA!—appending whatever information is readily available: habitat, preliminary description ("another beetle"), location, any local details. Then leave the rest to a future that will know its own capabilities and needs far better than we do.

To further the library analogy, would one rather have the card catalog or the books themselves? If they have comparable costs, the choice seems obvious.

MICROBIAL DIVERSITY

While "charismatic vertebrates" dominate the public consciousness of biodiversity, microbes underpin all the biosphere. Microbial diversity dwarfs the rest of the natural world; after all, they came first, and have had billions of years to fashion themselves.

Only a few thousand microorganisms have been described, but current estimates suggest a total species count of several hundred thousand to several million. They and the insects dominate the total world species count.

Microbes are not bit players in the global theater, but rather, integral actors in ecosystems. Dismissed as mere parasites even by biologists, their essential role is neglected. As biologist Norberto Palleroni has remarked, when the passenger pigeon vanished, none mourned the extinction of two louse species the pigeon bore.

Microbes tell us much about the strategies and limits of life, as they have developed for over three billion years. In the laboratory they are workhorses in the study of population biology. They are central to the biosphere, the workhorses of the planetary biogeochemical cycles. They carry the largest untapped reservoir of traits

KNOWN AND ESTIMATED NUMBERS OF
SELECTED BIOLOGICAL SPECIES

Group	Known Species	Estimated Total Number of Species	Percentage of Known Species
Mammals	4,170	-	near 100%
Birds	9,198	-	near 100%
Fishes	19,000	21,000	90%
Mollusks	50,000	-	-
Porifera	5,000	-	-
Crustacea	38,000	-	-
Insects	800,000	6–10,000,000	8–13%
Nematodes	15,000	500,000	3%
Protozoa	30,800	100,000	31%
Angiosperms	250,000	270,000	93%
Bryophytes	17,000	25,000	68%
Algae	40,000	60,000	67%
Fungi	69,000	1,500,000	5%
Bacteria	4,760	(>>)40,000	(<)12%

of value for biotechnology. They respond quickly, making them useful monitors of environmental change. They will play a role in restoration of a higher organism, for they comprise its vital background.

Yet we know little of microbial spatial patterns, and culture less than one percent of that visible in the microscope field of a soil or ocean sample. Indeed, we keep few systematic soil samples, though they have proved crucial in reintroducing plants to damaged environments.

A Library of Life strategy can readily respond to these facts. Only microbes are known to be fully recoverable from liquid nitrogen storage. They are easily gathered in soil or water samples, at varying depths, and compactly stored. Only one cell, not two, need be salvaged to save each microbial species (though for diversity we would prefer more). Given their probable future use in biotechnology, corporate support for their saving and storage may prove easy. Overall, microbes are the simplest form of biodiversity to preserve *ex situ*.

Given the importance of microbes, and their current neglect, a freezing strategy seems particularly appropriate. If all in situ measures fail for microbes, the Library of Life will still work for them, for they can be revived.

TROPICAL TACTICS

As an example of a truly massive program, consider sampling the rain forests.

This environment poses perhaps the most difficult, stratified problems. Ground-based sampling methods of terrestrial insects and animals are common and well-established. Litter-sifting techniques can be especially effective in capturing a large diversity of species. Tactically, this is ground well covered. But big problems loom overhead.

Rain forest trees have distinct stratification, with the tallest attaining heights of forty or fifty meters. Indeed, some forests in Southeast Asia can reach as high as seventy meters. Most photosynthetic activity occurs at the canopy, rich in leaves, fruits, and flowers. As well, the canopy hosts countless insects, air-dwelling plants, and many higher order animals.

This complicates study, as there is no convenient means for reaching the canopy. We have direct access only to the bottom two meters of the forest, or about five percent by volume. Unfortunately, these two meters are often the *least* productive zone. Clearly, gaining access to the canopy is paramount if we are to accurately sample the forests' diversity. Climbing trees to sample the uppermost layers is arduous, dangerous, and slow. Researchers have spent months climbing stands of fewer than a hundred trees, attempting to gain statistical information.

The Library of Life proposal does not seek this type of information. There would be no attempt to determine quantitative information such as species density or diversity. Furthermore, resources do not exist for thorough sampling of every plot on a global scale. Nevertheless, the canopy must be sampled.

A simple, field-tested method of sampling the canopy involves fogging it with biodegradable pyrethrin, a fatal fog. Many insects and other invertebrates fall onto plastic sheets to be collected. This method led biologist Terry Erwin to his 1988 estimate of thirty million insect species. It is also effective in collecting a great variety of species.

Other research focuses on accessing the canopy from the sky, via dirigibles or inflatable platforms that rest on the tops of the trees. These exotic methods are still experimental but should be discussed; relatively small, successful investment here could greatly simplify the collecting task and lower costs.

Now consider expense. After the initial investment for equipment and collecting, which could cost $100 million, the biggest expense will be replacing liquid nitrogen, a cost that eventually dominates. This depends ultimately on how much organic matter is collected.

Details of how big the sampled areas should be, how many plots should be sampled, and so forth, need statistical discussion. For concreteness, consider a simple method.

Suppose we sample 9,000 plots of about a hundred meters on a side. This is one for every thousand square kilometers of rain forest. If each plot costs $10,000 to sample (likely an overestimate), then the sampling demands $90 million. Assume one hundred trees are sampled per plot, taking ten kilograms from each tree and its surrounding area, including soil. This yields nine million kilograms to be preserved. The greatest bulk of sampled matter, by volume, will be plant material, with a fraction of the density of water. Assume this fraction to be one for the moment, and assign it to the packing fraction, discussed shortly. This means we need 9,000 cubic meters of cold storage.

A variety of biomedical freezers and storage vessels abound, but most can merely hold small vials for sample storage. By far the most efficient are the largest, so-called "big foot" containers. These stainless steel, cylindrical containers (called "dewars" for their inventor) are vacuum-sealed and internally wrapped 75 times with "super insulation." They have an effective capacity of 1,760 liters, with a liquid nitrogen boil-off rate of only twelve liters per day.

To allow for the wide range in packing methods, and sample densities ranging from tubular plants to topsoil, let us assume a volume packing fraction of 75 percent. Then we need about 6,800 "big foot" containers. With liquid nitrogen costs at current levels of about twenty cents per liter, the cost of keeping the samples frozen for a hundred years comes to about $600 million.

The "big foot" dewars can be mass produced at under $10,000

each, so containers will cost $68 million. The "up front" costs for sampling, containers, and a $20 million facility, come to $178 million. Even with $200 million for employee salaries and the $600 million for liquid nitrogen, it is difficult to imagine a total exceeding a billion dollars—an average cost of only $10 million per year over a century.

In contrast, to purchase several thousand square kilometers of the forest and hold them could easily cost several billion dollars over a century, even if political forces did not make this impossible.

Conservative estimates have been made throughout this estimate, so the total may be considerably less. The most uncertain variable is the average density of samples, multiplied by the packing fraction. Assuming we can cleverly pack materials at 0.75 the density of water may be optimistic. Since the largest term, nitrogen cost, varies inversely with this fraction, storage for a century could cost $6 billion if the fraction were 0.075, for example. Clearly this is the critical parameter.

However, accurate estimates are not crucial at this stage. This exercise shows that the idea is indeed affordable and quite inexpensive, compared with buying and maintaining a reserve of the sampled area in the tropics.

This simple scenario uses existing technology, undoubtedly not the best possible. For example, larger and more efficient dewars should be developed, for substantial savings. Also, liquid nitrogen accounts for the greatest expense and probably can be made at less expense in large, dedicated production plants. Lowering the cost of liquid nitrogen by even a few cents per liter yields huge savings; about $30 million saved for every penny decrease in the nitrogen price per liter in a century. This would easily justify spending a few million for liquid nitrogen production facilities.

However, this standard technology may deceive us; cheaper methods may soon emerge. In contrast to the original proposal in 1992, here I have not used the favorable surface/volume ratio as the collected mass increases. This reflects a simple physical fact: larger objects warm up slower than small ones. But a huge sphere filled with many thousands of samples seems unlikely, even if it would be cheaper to chill, because access would be difficult. Instead I have used available technology, though the maximum size is set by the largest dewar. This reflects the present state of the art, which steadily improves.

CAN WE AFFORD IT?

Such a sweeping proposal avoids the problem of deciding which species are of probable use to us, or are crucial to biodiversity. By sampling everything we can, we avoid some pitfalls of our present ignorance.

Many conservationists are reluctant to support a cryopreservation campaign, because they fear it will sample too sparsely. This assumes that present taxonomic methods and costs are necessary. But an important feature of this proposal is that the samples need not be studied as they are taken. This avoids the scarcity of taxonomists, speeding fieldwork and lowering costs. Plausibly, much of the gathering can be done with semiskilled labor.

This suggests immediately that the bulk of the funding come from "debt swap" between tropical and temperate nations, as has been used to "buy" rain forests and set them aside from cutting. Further, this will create a local work force that profits from controlled, legal forest work, rather than from cutting it. Current outstanding debt by tropical nations well exceeds a hundred billion dollars.

Of course, this does not touch upon side costs in training biologists, transport, perhaps doing some taxon discrimination, etc. Certainly the effort compares in cost with the Human Genome Project. The task is monumental; so is the plausible benefit.

Traditional economics cannot deal with transactions carried out between generations. As Harold Morowitz has remarked, the deep answer to "How much is a species worth?" is "What kind of world do you want to live in?"

COUNTERARGUMENTS

Published in 1992, this drastic proposal met immediate controversy, as I expected. I had deliberately framed an extreme form of the *ex situ* argument, and this was already a minority approach. I felt that the arguments about biodiversity had become ritual, with little progress. Thinking anew might help.

As well, I did not address the many legitimate reasons for preserving ecosystems intact, and I did not want the Library of Life to be seen as opposing them. Indeed, only by preserving *in vivo* a wide cross section of biota can we plausibly use much of the genetic library frozen *in vitro*.

Still, I was naive in the ways of science policy. In a letter to me, Carl Sagan wisely counseled:

> On the whole, I think it's an excellent idea. My main concern is that people will conclude that scientists have given up on preserving living biodiversity, or that future species extinctions are not so worrisome because we can always reconstitute the species and genera that we render extinct. But I agree that these potential obstacles can be circumvented: by stressing that only a fraction of the disappearing species would be "saved" this way and that the very fact that such steps are being taken is an indication of how serious the problem is.

Nonetheless, many prominent conservationists were unhappy. E. O. Wilson remarked to me that he thought such ideas should be carried further, with trial runs in the extinction zones we know, but in his authoritative *The Diversity of Life*, felt that:

> Cryopreservation is at best a last-ditch operation that might rescue a few select species and strains certain to die otherwise. It is far from the best way to save ecosystems and could easily fail. The need to put an entire community of organisms in liquid nitrogen would be tragic. Its enactment would be, in a particularly piercing sense of the word, obscene.

I knew I would provoke automatic disagreement with many in the conservation community. It was interesting to see this play out in the referee reports, a subject central to the scientific enterprise and yet seldom mentioned in public. Here I think they are instructive.

For example, *Nature* rejected the paper, their only stated reservations being ". . . it seems to us that your proposal is not practical, nor would it present a clear advantage over existing schemes or proposals." I had expected this; many biologists believe cryopreservation, since it involves technology, is pricey. "Not practical." Yet the principal founders of the San Diego "frozen zoo," Oliver

Ryder and Kurt Benirschke, had checked my calculations and supported this idea fully, believing it could be attained by scaling up their operations. Since I had received letters from luminaries such as Paul Ehrlich, E. O. Wilson, Thomas Eisner, and Michael Soule, all endorsing this idea as a major departure from present piecemeal efforts, I wondered who *Nature* had consulted. As UCI biologist Harold Koopowitz remarked, "When was the last time a biology proposal explicitly discussed costs *and* funding sources?"

The rest of the *Nature* criticism, that cryopreservation had no clear advantage over existing schemes or proposals, suggested that the editor did not know the literature, as one might expect of an editorial board that does not actually do any research. A similar result at *Science* convinced me that new ideas advanced by an outsider scientist, right or wrong, had rough sledding in the groves of establishment science. The in situ community is by far the largest and most vocal among the conservationist biologists, with strong emotional ties to their territories. Unsurprising, then, that they see any other proposals as attacks upon their funding. All too often, research is a zero-sum game.

No matter; once published by the National Academy, the paper attracted endorsements in an editorial in *The Economist*, notices in *Scientific American* and *New Scientist*, and many letters.

No one who proposes investment now—to send not merely a deep time message, but an entire library—can expect ready agreement. And, of course, to most this approach probably seemed simplistic and technologically far-fetched.

Still, there was enough interest to stimulate a two-day workshop in November of 1994, chaired by myself and Harold Koopowitz at the Beckman Center of the National Academy of Sciences, at the edge of my campus, the University of California at Irvine. Many subtleties came to light.

David Wildt of the National Museum of Natural History discussed the formation of a genome resources bank, designed to act as a storehouse of genetic diversity for endangered animal and plant species. As the total population within a species diminishes, so does genetic variability and its consequent protection against disease, and it experiences a loss of vigor in general; an organized effort to save the germ plasm and tissues of threatened organisms is insurance against catastrophe, disease, and political unrest.

Oliver Ryder of the San Diego Zoo pointed out that the solution to the biodiversity crisis—a "Biodiversity Ark"—is long-term and as much internationally political as technological. Ryder discussed the San Diego Zoo's "frozen zoo" of gametes and tissues from zoo specimens, then he distributed small cards containing spots of genomic DNA from a variety of endangered species as a concrete example of recent advances in technology.

Mark Martin of Occidental College discussed the need for understanding microbial diversity as a vital underpinning of overall ecological complexity. Practically nothing is known of the extent of microbial diversity, yet these communities of particular microbes could well be an important base upon which entire ecologies depend. Everyone noted how little is actually known regarding complex ecosystems. At present, despite so much labor in the field, we fail to capture the diversity of organisms.

Perhaps the greatest asset of a broad, random sampling would be to capture the diversity of organisms, from the microscopic to the towering, all of which provide a support system for such threatened animals as pandas or Siberian tigers. Such a support system may be highly resilient to habitat destruction and environmental degradation, or various key members of such communities may be exquisitely sensitive.

Obviously, a multidisciplinary approach, bringing together microbiologists, geneticists, ecologists, conservation biologists, economists, industrialists, politicians—and a leavening of women and men from other disciplines—will be necessary to address such a large and important task.

Of course, much of this will be unpopular. Compared with saving tigers or graceful birds, sampling and freezing have little aesthetic appeal. Let us be frank: we get no immediate benefit, yet bear much of the cost. This does not play well to an audience that expects pandas on display at nearby cages in return for a zoo membership.

To some, cryopreservation smacks of fatalism; it may be merely realism. As well, freezing species does not offer the immediate benefits that preservation yields. (Samples would probably be taken only from areas not already highly damaged.)

More concretely, this proposal will not hasten benefits from new foods, medicines, or industrial goods. It will not alter the essential services an ecosystem provides to maintenance of the biosphere. We should make very clear that this task is explicitly designed to benefit humanity as a whole, once our age of rampant species extinction is over.

Some will see in this idea a slippery slope: to undertake salvaging operations weakens arguments for biodiversity preservation, as Sagan pointed out. There is no intrinsic reason why this need be so. They are not logically part of a zero-sum game, because they yield different benefits over different time scales. Of course we would all prefer a world that preserves everything. But the emotional appeal of preservation should not disguise the simple fact that we are losing the battle, or to argue against a prudent suspension strategy.

To avoid this argument, the two *parallel* programs of preservation and freezing must be kept clear. In this sense the analogy to the library at Alexandria is false—for us, there is no true conflict between fighting the fire and salvaging texts. Further, in the real world, funds for conservation of DNA today do not come directly from in situ programs. If the Topeka Zoo budget is cut, the city does not transfer the funds to Zaire to save gorillas.

Indeed, one can make the opposite argument—that the spectacle of the scientific community starting a sampling program will powerfully illuminate the calamity we face, alerting the world, stimulating other actions.

Beginning with local volunteer labor and contributions—say, with the Sierra Club sampling the redwood habitat—could generate grass-roots momentum to overcome the familiar government inertia. In larger campaigns, by requiring that samplers accompany all legal logging operations, we can help develop a local constituency for controlled harvesting. Further, sampling is far less expensive than preservation—which is why it is more likely to succeed over the long run. Even competition for "debt swap" funds will not necessarily be of the same economic kind. Conservationists seek to buy land and set up reserves, putting funds into the hands of (often wealthy) landowners. A freezing program will more strongly spur local, largely unskilled employment, affecting a different economic faction for at least the duration of the sampling.

Still, the most difficult argument to counter is basically an unspoken attitude. As scientists we are trained to be careful, scrupulous of overstating our results, wary of speculation—yet these militate against the talents needed to contemplate and prepare for a future that can be qualitatively different from our concrete present. Paradoxically, by our labors we scientists often bring about this changed future. It seemed to me that now was the time to bank on the expectation that we will probably succeed.

A BIOLOGICAL OZYMANDIAS

Sadly, there is no quick and easy fix to the coming calamity the world faces. Most of the remaining tropical forests probably will be damaged or destroyed. We should now view this not as speculation, but a matter of near fact. Clearly, it is time to examine comprehensive, long-term conservation strategies that integrate both in situ and *ex situ* aspects.

I was quite aware that I was an amateur stepping on a lot of important toes. But given that major figures in biodiversity argue that a large scale species dieback seems inevitable, leading to a blighted world that will eventually learn the price of such folly, it seemed that I had to propose something radical, in keeping with my experience in deep time messages.

What are the odds for a Library of Life? Small, as viewed from the perspective of 1998, but I believe opinion will shift as the tide of loss grows. The National Academy workshop galvanized some workers, heightening concern. I suspect another decade of damage must pass before the extremity of our situation penetrates. With our population projected to increase from the present less than six billion to nine or ten billion, all within fifty years, the drama will play out in slow, inexorable motion.

In the long run, social impacts of biodiversity loss will be immense. Politics comes and goes, but extinction is forever. A less diverse natural world is more fragile and offers fewer happy discoveries. We may be judged harshly by our grandchildren, our era labeled the Great Dying or the Age of Appetite.

A future generation could well reach out for means to recover their lost biological heritage. If scientific progress has followed the paths many envision today, they will have the means to perform seeming miracles. They will have developed ethical and social mechanisms we cannot guess, but we can prepare now the broad outlines of a recovery strategy, simply by banking biological information.

Such measures should be debated, not merely by biologists, but by the entire scientific community and beyond, for all our children will be affected.

These are the crucial years for us to act, as the Library of Life burns furiously around us, throughout the world.

PART FOUR

STEWARDS
OF THE EARTH:
THE WORLD AS
MESSAGE

The land was not willed to you by your ancestors—it was loaned to you by your children.

                       —KENYAN SAYING

We have inadvertently sent down through history a crucial deep time message: the world itself, as we have made it. Our planet, the background for our great dramas, is itself a mixed message from antiquity, shaping all our unconscious assumptions.

We have been stewards of large parts of our planet for quite some time—and rather poor ones, as a whole. What we now regard as a pristine landscape was actually fashioned by perhaps a hundred thousand years of our forebears' intervention.

Our present view of the natural world is a fiction. Western society did not inherit a wild, pristine world, either in Europe or in ventures overseas. Paleolithic communities around the world set fires to shape their land. Hunter-gatherers used fire at least sixty thousand years ago in Africa. For millennia tribes heavily managed the Serengeti plains, burning grass to provoke new growth and pruning herds. Slash-and-burn farming already thrived in Costa Rica and Panama six thousand years ago, lands thought until recently to have never been "spoiled." Human-set fires shaped both the African savannah and the American prairie. Rain forests in Latin America and Polynesia boast many plants brought by the first humans from Asia.

Many cultures pillaged resources, downed forests, and sterilized fields with unwanted salt—from the first civilizations of the Tigris and Euphrates rivers (now in Iraq), through the Mayans and on to our present.

Archeology's greatest contribution may well be to illuminate a central, sobering fact: we have been sloppy stewards for a long time. Many early societies collapsed from erosion and soil degradation. These were not the wise, benign cultures beloved in romantic legend, living in balance with their surroundings.

A principal tool for seeing so far into the past is the study of ancient lake sediments. Silt collects there, layered lessons in the fate of early agriculture. Pollen sediments tell us when trees were felled, what crops grew, and the rates of soil erosion. When wood ash disappears, one infers that the forests were gone. Often there follow the clues of animal dung ash, the ancestors of energy-starved practices used in India today.

Archeologists know the signatures of such ancient deforestations well. Telltale ash from burning buildings marks warfare, usually the last act in environmental collapse. Such dark scenarios had already played out around the globe in ancient times, though this is a recent discovery.

Americans have a popular illusion about our continent, which we suppose was thickly wooded and wild when we arrived. In fact, the earliest explorers remarked on the extent of grasslands, and that forests had little undergrowth, so that, as Captain John Smith of the Jamestown, Virginia, colony remarked, "A man may gallop a horse amongst these woods any waie, but where creekes or Rivers shall hinder."

Virginia wasn't virginal. The land reminded Europeans of their well-tended gardens, for good reason. The Indians had regularly set fire to the woods. Universally, they wanted to clear underbrush so that enemies could not approach unseen. This also helped hunting, for a deer could be seen a mile away in a cleared forest. John Smith described how Indians killed deer from canoes after they had been driven by fire onto a peninsula. Hunters routinely selected and culled game, except for those animals who could evade us or defend themselves. Hills were torched to send signals to gather tribes, or to control the battlefield in warfare, and to clear ground for crops. This left a mosaic of relatively young trees and grasslands, so that forests were younger before Europeans came than is the case in parts of North America today.

New England forests were burned roughly every decade to clear them for movement and hunting, the hickory and oak survivors making a beautiful forest: nature arising from artifice.

Yet the modern mind faces this truth reluctantly. We prefer a pristine past. Even this impulse is old: ancient Greece and Rome routinely (and incorrectly) invoked earlier, better times when civilization was higher, more virtuous. Indeed, we accommodate ourselves to the consequences of ancient depredations, seldom sensing these as clues to the true history of our surroundings.

THE UNSEEN HERITAGE

In 1997, I had a glimpse into how adroitly humans adjust to their own alterations of the world. Outside Calcutta, I visited a Hare Krishna compound that had an extensive, neat corral and a system of concrete chutes to carry cow dung into an underground dome. There it decays and methane collects above it. The dome's concrete blister pokes above the ground next to the collective kitchen. From the dome's pressure trap they pipe methane directly into the stoves and ovens upon which the cooks were preparing lunch. The gas burned cleanly, blue and hot. A touch of Western engineering plus a plentiful resource.

How clean was it? I asked. Very, they said. This was certainly a better solution than the traditional, I thought, which we saw along many plaster walls and the sides of houses: dung pancakes shaped by hand and stuck up to dry. When they are ripe, they fall off, are collected and stacked in ricks to dry further. Eventually they are burned in the huts for heat and cooking fuel, the smoke inhaled by the whole family.

The method yields a distinct flavor, communicated to the food. Later, among the neatly arranged, traditional, thatch-roofed classrooms I saw two women spreading a brown layer on the dusty paths. One mixed cow dung in a bucket of water and the other smoothed it everywhere with a broom. Why? Ancient Vedantic lore holds that cow waste is medicinal. It sanitizes the area, my guide said; much cheaper than Western products.

The desire to make do with natural substances runs deep. The earnest attendants lauded each application as "natural," in balance with nature. None realized that it was a fallback adopted in antiquity when the Indian forests ran out.

Islands reveal such patterns even more clearly. Their constricted ecologies decline faster under stress and are less forgiving. The

Norse colonized offshore islands brutally—a pattern of rape, ruin, and starve. They overpopulated the islands, overgrazed the fields and eroded away the already thin topsoil. The Norse were barely clinging to their islands when the Little Ice Age of the 1370s brought their death knell. Those in Greenland and Vinland died out, with Iceland barely holding on.

On Easter Island, within a few centuries the immigrant humans depleted the forests and many game species. Fierce warfare broke out until there was precious little left for anyone. This was a particularly bad case, though a pattern of quick extinctions appears on islands throughout Polynesia.

A large lesson of antiquity is that farmland deteriorates when controlled by central government, just as in present-day China and Russia and some African socialist states. Local farmers usually know better how to maintain soils. Distant rulers demanding ever more grain as the cities swelled paid little attention to the small voices from below. But until quite recently, the more subtle dangers escaped even the men and women in the fields.

Irrigation water, at first a seeming boon, leaves mineral salts behind when it evaporates. Mesopotamia learned this too late. In 2000 B.C., Ur was the very model of civilization, with writing, public monuments, and prosperous winter wheat fields. As salt built up, the rulers decreed a shift to the more resistant barley. By 1700 B.C. barley had vanished and grain was imported, until overpopulation brought on classic city-city warfare over food stores, with starving citizens crouched behind heavy walled defenses. One of the last written records from Ur is a plaintive message from the king to a functionary in the hinterlands: "Where is my shipment of grain?"

The legendary Arabian peninsula city of Ubar suffered a literal collapse. Around A.D. 500 it fell into a sinkhole, probably because it had depleted the aquifer beneath it.

Of the Mesopotamian catastrophe, archeologist Charles Redman of Arizona State University remarked, "These were a gifted, intelligent people, the pinnacle of the past. If they couldn't handle their interactions with their environment, that is a serious condemnation of human organization."

The Mesopotamian lands lie barren, deadened by salt. Similar salt deserts stretch across what was once the breadbasket of the Roman Empire. Salting up also doomed the Hohokam who lived in Arizona from about the time of Christ until A.D. 1400. Their elaborate irrigation system salted the land, which has still not re-covered.

Present dwellers in Phoenix believe they are fighting the desert's natural problems, but in fact their landscape is an artifact of a mil-lennium ago. The Chaco Canyon area in New Mexico was once magnificently forested. From it the Anasazi built five-story build-ings with forty-foot rooms spanned by timbers from huge trees. There is not a tall tree in the entire area today. Their complex society simply vanished, fallen by its own systematic deforestation. A drought accelerated the decline, but in the judgment of most archeologists was not the primary cause.

Similarly, the great Central American societies of the Aztecs and Maya flourished for more than a millennium and then, one by one, suffered collapses of varying severity. The Maya had destroyed over eighty percent of their region's forests by the end, around A.D. 900.

When the Spanish invaded, the Maya had shrunk to a quarter of their peak population, barely holding on to eroded lands destroyed by slash-and-burn farming that was pushed to a hundred-mile pe-rimeter around the major cities. Ironically, the Spanish brought diseases which laid waste to the people but ultimately revived the environment. The rain forests still growing in the region are a grad-ual return of the original environment.

North America was much like the Serengeti plain of Africa when humans arrived, at least twelve thousand years ago. Many of the easy game animals then native to North America were eliminated: mammoth, stagmoose, camel, musk ox, and more.

Agriculture did far more, reshaping whole river valleys. Farming casts long shadows through history, spreading weeds and contour-ing hills. Worldwide, exotic species mingle with natives, creating biosystems that never would have occurred but for us. Mice and rats flourish because we store grain and discard garbage.

Most immigrant species blend in; the destruction caused by brown tree snakes in Polynesia, kudzu in the U.S. South and burros in the West, are atypical. Most traveling species are inherent col-laborators. Indeed, weeds get their start by being wily adapters, moving into the disturbed earth humans make as we uproot, plow and build.

As a species we have served for at least a hundred millennia as social directors for evolution, introducing species to each other.

Carrying them in our seed grain, or just stuck on our boot heels, we have carried seeds of new species into distant lands. There we gave them fresh opportunity by providing large stretches of disturbed ground, a talent unique to our species.

Slowly we are learning such intricacies of our biological past. Such knowledge is not of merely academic interest. Archeology proves crucial in demonstrating how large a population a region could support, and the clues warning of impending collapse.

Though our understanding of farming is far better now, studying such large effects is difficult with short-term data of a few decades or even a century. Archeology can provide a millennium of soil records, and some written human reactions to gathering crisis. This will prove useful in the twenty-first century, when we will have to make clear-eyed judgments of how much human burden a given region can support. In planning biological preserves, this will be vital.

But the past is only a qualitative guide, for our mastery of matter looms far larger now. The perspectives of deep time can instruct, but the regional catastrophes of antiquity pale beside the damage we could do in the centuries to come.

A PLANETARY PARK

Millennia of human "meddling" have all but erased the meaning of the word "natural."

The largest remaining region bearing some semblance of its distant past is probably Siberia, which has benefited from the historical incompetence of Russian rule; under better management, it probably would be more developed. Even so, widespread hunting eliminated many species that once were safe from us in the long, mosquito-infested valleys of central Asia.

Intervention, often unintentional, appears everywhere in the landscapes we think to be "wild." Nature is not just a series of landscapes, but a process. We now dominate many of its workings. There are many more trees in southern California now than two centuries ago, but most are recent immigrants such as eucalyptus.

Often such changes make the land seem more beautiful. We prefer to live at the edges of environment, amid grasslands and forest close beside bodies of water. It is no accident that golf courses look this way, admirably compact editions of the savannah playing to our preferences.

This does not mean that we have intervened wisely, of course. The deep time message sent forward by our ancestors is that of neglect, based upon assuming that the world is essentially infinite, forgiving of our indulgences.

We moderns can take no comfort in our own acts, however. At present we are sending on a message of disrespect for both the Earth and for our descendants. For the first time, we are affecting the entire globe in a unified way.

Only reluctantly do we see our world as artificial, in part because much of it is beautiful, which some current thinking holds incompatible with human interference. Yet much "natural" beauty comes from our efforts: managing forests, grasslands, and waterways. Typically we see ourselves as sinners against nature since the Industrial Revolution, while we presume ancient socieities were virtuous.

Bill McKibben in *The End of Nature* contrasts our time with the distant past: "We have built a greenhouse, a *human creation*, where once there bloomed a sweet and wild garden." But gardens are human creations, too, funneling natural processes along avenues ordained by the gardener. Grafting, after all, is quite unnatural. It has yielded roses and cherries and tulips that would never emerge from nature. Yet we seldom think of grafting because it is an old technology, the first way we shaped life to procreate without sex.

As Daniel Botkin remarks, surveying our true natural history, "Nature in the twenty-first century will be a nature that we make; the question is the degree to which this molding will be intentional or unintentional, desirable or undesirable. Having altered nature with our technology, we must depend on technology to see us through to solutions." For Botkin and other "new ecologists," nature has no "true" state, contains no moral imperatives in and of themselves.

Stewardship is a view inherited from feudal times, an era in which social hierarchy dictated policy. Such a social model obviously will not work today. What we need in the long run is a new theory of social and economic accounting, one without even theoretical underpinnings as yet.

We all know that we face major environmental problems in the next century. To pass down to our distant descendants a planet of adequate resources means that we must manage it now. In this sense, the greatest deep time message we shall send is the condition of that future, admittedly artificial, world. We can hope that centuries from now Earth will shine like a blue-white jewel in the solar system, still the finest place to live—but like a jewel, it will rest in a setting of embracing artifice.

FIXING THE GREENHOUSE

Only by acknowledging that our world is no longer "natural" can we then embrace artifice as inevitable. As an example of what deep time message our era could send down through centuries, I shall concentrate on one area of stewardship which, mishandled, may obscure all others.

Forty years ago noted oceanographer Roger Revelle declared that "human beings are now carrying out a large scale geophysical experiment"—yearly pumping billions of tons of carbon dioxide into the air. We have taken this long to get serious about the issue of inadvertently "geoengineering" our planet by altering its atmospheric chemistry.

No issue holds more profound possible consequences for the next century. Yet so far the debate and hand-wringing have been both angry and unimaginative. There may very well be fairly simple ways, and even easy ones, to fix our dilemma—but the tone of discussion never makes this clear. Most proposed solutions are a funhouse mirror, telling us more about our moral postures than our complex world. Debate swirls over evidence, transfixed by details, largely ignoring the horizons.

Some facts are incontrovertible. Carbon dioxide (CO_2) levels have risen thirty percent since the Industrial Revolution, and the planet has warmed about half a degree centigrade in the last century. Tiny air bubbles trapped in glacial ice show that past climate variations closely followed CO_2 levels, now at the highest in 200,000 years. They also reveal that sudden shifts like the last century's have been rare, appearing only once or twice a millennium. Spring arrives about a week early now in the arctic tundra. In Antarctica, spring flowers enjoy a growing season two weeks longer than twenty-five years ago. Ice shelves retreat near the poles. Marine fauna near the shores are extending their range to higher latitudes.

Similar plant growth patterns imply, but do not prove, global warming.

Still, atmospheric specialists nearly unanimously agree that changing global patterns lie behind a rapidly growing body of suggestive evidence. We know only enough to plausibly foretell that climate may change from our effects; "global warming" is a catchall for this.

Earth's greenhouse retains the sun's heat, delivered by visible light, because certain gases trap low-frequency heat radiation, not allowing it to escape to space. Carbon dioxide and several other gases are very efficient at doing this. I find the evidence of this global trend convincing, and shall assume so here.

Climate changes are humbling, reflecting how poorly we understand the entire planetary thermostat. Some broad facts are clear: from the sun comes 1340 Watts per square meter (W/m^2), striking the top of our atmosphere. About 100 W/m^2 gets reflected, 80 W/m^2 are absorbed within the atmosphere and 160 more absorbed by seas and land. Some of this soaked-up energy returns to space, carried out by invisible heat radiation, the infrared. Water vapor absorbs some, rises, and creates weather.

Clear enough, but then complications stack to the sky. Water clouds both reflect sunlight and absorb infrared, just how much depending on how thick they are and on the height of their tops. Natural water vapor may be the most important greenhouse gas, but we aren't even sure of that. Hopes that warming will give us cloudier days, reflecting more sunlight, may prove true—but are controversial. Pollution clouds of sulfate particles over cities do reflect sunlight, and so have partially offset the usual greenhouse effect of burning the fossil fuels that made the impurities in the first place. The first effect of reducing emissions will be to lose that reflecting layer, contributing further to warming. Climate study is a blizzard of such details; every coin has two faces.

Just how complex the air-ocean-land structure may be is itself controversial. At the extreme lies the Gaia hypothesis, which envisions a self-organizing and regulating entity that has adjusted over billions of years to keep Earth's biosphere vital in the face of astronomy's blunt forces and steadily building irritants, such as salinity in seawater. Most biologists and geologists reject Gaia but concede that we only dimly perceive how the system works. Variations may be sensitive to seemingly minor effects.   •

For example, in the early 1990s a big good-news discovery showed that atmospheric methane concentrations have stopped rising. This cheered many, since methane is a potent greenhouse gas. There are myriad sources of this gas, which accounts for a quarter as much warming as CO_2. Presumably it comes from leaky Russian pipelines, termite digestion, rice paddies, cow flatulence, swamp gas, and the like. Yet there have been parallel *drops* in CO_2, while oxygen content jumped. How this fits with the methane plateau remains mysterious, especially since the latest data shows levels resuming their rise again. The blunt fact, hard for us with our short attention spans to swallow, is that the entire biosphere reacts sluggishly on a scale of a century or so, taking that long to scrub gases such as carbon dioxide from the air.

No climate modeler pretends to a detailed, quantitative understanding of what's going on. Yet we may be rushing toward an era when skimpy knowledge will be no excuse to not act.

GREENHOUSE GASING

Mounting evidence led the U.S. to sign the Rio Earth Summit pact in 1992, promising skeptics we would reduce greenhouse gas emissions to 1990 levels—by 2000, honest. All this they would do by relying on voluntary measures—honest.

But by mid-1997 the State Department owned up that we are now eight percent over 1990 levels and would miss the target bigtime, venting thirteen percent more than in 1990—and some watchdog groups think that's an underestimate. Meanwhile, scientists' estimate of the correct target to avoid big warming effects is about forty percent less than the 1990 levels.

Economists, not fond of making moral judgments, blame our burgeoning economy on the growing use of fossil fuels. At the State Department there is talk of asking to push the date back to 2010, when better technology might make a difference. They have heeded some startling estimates that the industrial cost to drop U.S. CO_2 emissions by twenty percent could be several *trillion* dollars. As usual, there are economists who differ enormously about this, leaving the layman scratching his head.

Certainly the easiest technical way to curtail CO_2 lies in automobile fuel economy. Congress has ducked this path, ever since it trampled President Clinton's 1993 energy tax ideas. We are the big bad boy of global warming, four percent of the world's people using a quarter of the fossil fuels, but with plenty of company; most of the advanced nations will fail to reach their goals, too. There is plenty of blame to go around.

Planners propose several mechanisms to help comply with the Rio goals, such as carbon taxes and trading of emission credits. In 1997 nearly two thousand U.S. economists signed a statement arguing that the benefits of acting outweighed the costs, envisioning principally cleaner fuels, cars, and technologies. They advocate "recycling" proceeds from a carbon tax into lower payroll and corporate taxes, to stimulate new investment. Still, years of negotiation have failed to make any headway on these measures.

To answer a rising howl of international complaint, the U.S. proposed in 1997 a foreign aid program to end tree-cutting and burning policies in the *developing* nations, plus some technical aid. Catcalls greeted this; speaking at the U.N., Clinton avoided setting a new conservation target and said nothing that would provoke domestic consumers.

To be sure, the prosperous states have plenty of promising and nearly painless new ways to cut back: high efficiency refrigerators and lighting, ozone laundering, microwave drying, variable-speed motors. The microlevel tinkerers have no end of tricks. But so far even humdrum methods like better insulation, smaller cars, wind and solar sources, remain underused.

The six-hundred-pound gorilla in global warming is the rise in fossil fuel burning in the developing nations. The generally reliable International Energy Agency projects that 85 percent of rising CO_2 emissions will come from developing regions and Eastern Europe, societies with little appetite for conservation.

Projections show China increasing its burning by a factor of five in the next thirty years, when they will be using half of what the whole world uses now. China and Russia have immense reserves of coal, the worst polluter, and are moving to exploit them further. They have customers already standing in line.

China is an instructive case. Throughout the last decade its Gross Domestic Product has grown at eleven percent per year. Such surges in the underdeveloped world cannot conceivably be offset by such reductions as those envisioned in the Kyoto Protocol of 1997. Nor does there seem any incentive to change this situation among the poorer nations. As a native of the world's champion consumer society, I find it difficult to frame a moral argument against such growth.

New factors threaten, as well. Recently, oceanographers have found a potential new fossil fuel source: methane hydrates, condensed natural gas trapped in crystalline cages of ice. Seabed prospecting has found vast reserves on the continental shelf, reachable by many nations. Estimates say that methane hydrates hold double the combined reserves of oil, coal, and ordinary natural gas.

Even Draconian restrictions by the fossil fuel exporters—which seems a fantasy—could simply stimulate hydrate use. Fossil fuels are just too plentiful to regulate globally.

Climate experts doubt that we can avoid a doubling in greenhouse gas emissions. Some even foresee a tripling or quadrupling within fifty years. The global economic and political system simply cannot change rapidly. Costs of replacing present equipment with more energy-efficient or gas-trapping technology are considerable. All our best intentions cannot turn a supertanker in a sea of syrup.

OPTIMISTS ON CALL

Early in the greenhouse debate, some saw more good than harm. Could more CO_2 and warming help? After all, that's how nature took us out of the glacial age, propelling Cro-Magnon toward his current undreamed plenty.

Slight warming does indeed prod trees and crops to higher yields. This could help feed the coming population doubling some foresee within the next fifty to seventy-five years. But studies of the last century's warming effects on northern trees shows a plateau effect, beyond a mild heating boundary already crossed. With good use of added fertilizer, farmers might get a rise in yield, but perhaps for only a few decades before the effect ends or even reverses. Crops show some vulnerability to high temperatures, as well, especially without lots of added water.

In the early 1980s the energy companies launched a tobacco-industry style disinformation campaign against the global warming findings. Some scientists cast a skeptical eye at the data, while others were professional optimists. Predictably, the effort to "reposition global warming as theory rather than fact," as one fossil fuel lobby memo put it, spurred each side to higher ramparts of rhetoric.

Most media are vulnerable to the binary model of disagreement, so that the skeptic position on warming gets equal exposure, despite being a tiny minority of the scientists working in the area. This does not mean they are wrong, but it does reveal a sobering truth: a small propaganda investment by the oil and coal lobby has bought them decades of delay.

"We have no spare decades left," Bill McKibben declares, as economists agree that the U.S. shift from heavy industry to a service economy spawned few cuts in consumption. In fact, demand for the highest quality energy—electrical—swells as phone lines and computers proliferate. And economist Arthur Rypinski of the Department of Energy notes that "even in the information age it gets cold in the winter and hot in the summer."

Bad effects loom as far more probable in the long run. Pumping more energy into the weather system will alter patterns, perhaps making them more violent. Weather changes could give us crop-blighting droughts in Kansas, dustbowls in Asia, brutal hurricanes. Water expands when warmed, so sea levels could rise a foot or two, inundating farmland and many cities. Melting polar ice caps will add to this.

Oil companies aren't the only megabusinesses concerned with warming. Six of the ten most costly natural disasters in the U.S. occurred in 1986-96, bringing some insurance companies teetering toward insolvency. They have backed studies to see if warming is the culprit.

As well, the Alliance of Small Island States has lobbied for the advanced nations to reduce their greenhouse emissions twenty percent below 1990 levels. They fear flooding within mere decades, erasing whole island countries. Biologists, already wringing their hands over many shifts in plant growth that may signal global warming, need only look a bit downstream to see possible synergistic effects. Insects could migrate into higher latitudes, bringing tropical diseases to a new audience. Public health spending is an even touchier subject than insurance.

BECOMING STEWARDS

Some climate scientists worry that we may be approaching a chaos boundary, invoking ideas from current nonlinear systems theory. If so, the world could lurch suddenly into a dynamic equilibrium unlike the mild conditions civilization has enjoyed since the last glacial era, over ten thousand years ago. Geological evidence shows that over the last seventy thousand years the planet has snapped into severely different temperature regimes, for only vaguely fathomed reasons.

This has already prompted some to work out possible, dramatic shifts. A recent study showed that some northern forests were withering under a two-degree-Centigrade local increase over the last century. Slight warming, good; more, bad.

Following this trend further, if high latitude tundra melts, it could release stored methane, which is twenty times more effective at greenhousing than CO_2. Such a triggered shift at the polar regions could rearrange our global weather, damaging crops. Melting of the West Antarctic ice sheet could make the sea level rise several meters within a few years.

Should we worry? Chaos theory itself is more than a bit precarious. This young discipline applies to systems with few active degrees of freedom. So heavily constrained, they act effectively as if they resembled simple one- or two-dimensional dynamical systems, such as a pendulum that has a large arc.

Are such rarefied studies useful for the greenhouse? Quite possibly not: Climate has at least several major active factors—air, sea, and land, plus the sun, and doubtless other variables only vaguely glimpsed. With many degrees of freedom, one needs complexity theory—chaos theory for grown-ups. Alas, we know little of that currently hot area.

So in this crisis, we are stuck with inadequate theoretical tools. The history of science shows that good advice in such situations is to stick close to the phenomena, relying on small, weak experiments, well-tracked by computer simulations and an army of thinkers. This the advanced nations are doing, and being rather quiet about it. But they may not have long to labor.

Below the rarefied realms of theory, in the muddy media battleground, the hired optimists are losing the propaganda battle to the doomsayers. Disaster is more thrilling, makes for better graphics, and there are many varieties to choose from.

Noted climate authority James McCarthy of Harvard observed, "If the last 150 years had been marked by the kind of climate instability we are now seeing, the world would never have been able to support its present population of five billion people."

Economists tend to forget that the global industrial engine depends on the global environment, not the other way around. What happens when the business community starts projecting climate shifts in its plans? Alarm could spread, provoking corporate pressure to *do something*. But what?

"The only way to slow climate change is to use less fuel," McKibben asserts, echoing the universal environmentalist position. Indeed, ecologists and many other scientists champion extreme conservation measures as the only solution. Both scientists and environmentalists have long histories of distrusting the unruly market and the vagaries of diplomacy. Ross Gelbspan's *The Heat Is On* even urges a public takeover of the energy sector and a massive propaganda campaign. Expect to see such calls for a Greenhouse Czar as the problem worsens and rises to broad, persistent public notice.

In his last book, *Billions and Billions*, Carl Sagan stated flatly, "The world must cut its dependence on fossil fuels by more than half." In context he was envisioning Draconian police measures and a sweeping indoctrination campaign.

Those pushing better technology and solar energy, such as Michael Oppenheimer of the Environmental Defense Fund, do sometimes believe that goals "would be better achieved using incentives and disincentives than technology mandates and confiscation."

Still, they all see no way out but the Puritan Program: abstain, sinner!

POLITICS AND PARASOLS

A little noticed 1992 National Academy of Sciences panel report clarified the muddy science behind global warming and then ventured further. Could we intervene to offset the warming? Accept that greenhouse gases will rise and find ways to compensate for them? Mitigate, not prohibit?

In this view, our ongoing, inadvertent experiment would not be stopped, since that would condemn most of the world's population to a slower climb out of their poverty. The question should be how best to *design* our global experiment, to maximize benefits and minimize costs.

As the larger public in the advanced nations becomes convinced that global warming is an immediate threat, worthy of response, they shall ask for solutions that command the least sacrifice. Why should so large and powerful a fraction of humanity not act to maximize their short-term interests by minimizing economic and social inconvenience? They always have before. To save our world, then, we must modify its response to the gases we now pump into its air.

Climate modification is time-honored, though not clearly a winner. Cloud seeding in the U.S. during the 1940s and 1950s met some success, but ended in a blizzard of lawsuits from those who claimed their local rainfall had been coopted by neighboring areas. Though such assertions had little scientific proof, courts felt otherwise.

During the cold war both sides studied a menu of climatic dirty tricks. Plans to drop dark dust on polar snows envisioned also diverting or forming rivers. Think tanks contemplated how to bring crop-killing changes to opponents. Increasing cloud cover to put a parasol over an opponents' cropland just before prime harvest could distort economies. Apparently none got carried out, though there were lesser biological warfare measures aimed at Cuba's agriculture.

These programs floundered on a fundamental fact: before modifying climate, first one must grasp it. At the level of understanding of even the 1960s, only spectacular interventions would have left discernible signatures. Climate variability was so little fathomed that weather prediction was pointless beyond roughly a week.

Since then, in an advance little noticed by the public, systematic weather prediction has advanced more than tenfold in assured time range. By watching the sun, atmosphere, ocean, land, and clouds using satellites, advanced aircraft, ships, and a tightly gridded land-observing system, we have diminished the classic uncertainties in the long range weather. As Mark Twain pointed out, we are still just talking about the weather, but at least the talk is of higher quality and we can see a bit downstream.

In 1997 the U.S. National Oceanic and Atmospheric Agency

predicted a coming wet winter half a year in advance, based on temperature measurements of tropical waters, presaging a new El Niño. They proved right. This signaled a new era in forecasting. With the latest systems, backed by heavy computer modeling, we shall with rising assurance shrink uncertainties, identify subtle feedback loops, sniff out regional pollution patterns, discern the spread of deserts and the withering of forests.

Sensitive global measures of disturbance shall open further to us: polar cap and glacial contractions, ozone levels, volcanic dust, levels of the oceans. There is even a technique available for cheaply gauging the global reflectivity, by measuring "earthshine"—the faint glow of our reflected light, seen on the dark portion of a crescent moon. Using a small telescope and makeshift gear, astronomers easily showed that we reflect thirty percent of incoming sunlight back into space—a number that our satellite system got earlier, at a price tag of hundreds of millions of dollars. Such innovation will lessen the costs and confusions of global understanding, a help we shall need dearly as, and if, the greenhouse predicament worsens.

One way to think of global warming is that we are unknowingly acting on a time scale to which our global climate now responds sluggishly, taking about a century to manifest large shifts. Mother Nature takes her time, looking longer than the life span of individuals. We are mayflies.

Any correction by technological intrusion will have to occur on far shorter ranges. Most politicians consider a few years the far horizon. Statesmen look longer, perhaps a decade. Getting nations to think over a century scale will be a principal challenge of the new millennium. At least for the first time, we shall be able to use an armada of diagnostics to discern effects, their time scales, and perhaps traces of the causal chain.

GEOENGINEERING

Some engineered systems appear possible to deploy now, and at reasonable cost. They could be turned on and off quickly, if we get unintended effects.

Nobody envisions quick, full-scale climate mitigation. At first we should do laboratory work, carefully testing the basics of any proposed scheme. Then we can follow with small-scale field experiments to answer questions about how our current atmosphere behaves when one alters the kind of dust or aerosols in it. The biosphere is a highly nonlinear system, one that has experienced climatic lurches before (glaciation, droughts) and can go into unstable modes, too.

Some argue that this simple fact precludes our tinkering with "the only Earth we have." Earth's climate might be chaotically unstable, so that a state with only slightly different beginning conditions would evolve to a markedly altered state. The alighting of a single butterfly might change our future. But we also know that the Earth suffers natural injections of dust and aerosols from volcanoes, so probably, experiments which affect the planet within this range of natural variability should be allowed.

Still, suppose a big volcano erupts while you are floating artificial dust high in the stratosphere—might this plunge us into a new ice age, pronto? The proper answer is: not if we keep our artificial dust well below the historical fluctuation rate; and without experiments, we cannot make progress.

Global warming is assessed in a rather tightly knit community of scientists, mostly academic. They only study nature, while engineers dream of altering it. Both need experiments to guide them.

This portends the inevitable emergence of a new technovisionary community, devoted to solving global ills with global technologies. This is quite different from simply finding the polluters and forcing them to stop. Such good/bad dramas are an old theme in environmental issues; like depletion of the ozone layer and cleansing of the seas, global warming has provoked an automatic politicizing of any proposed solutions at birth.

With ozone and the seas, we could comfortably point fingers at big companies. Alas, the culprit in global warming is plain old us.

FORWARD TO THE FORESTS

The simplest way to remove carbon dioxide from the air is to grow plants—preferably trees, since they tie up carbon in cellulose, meaning it will not return to the air within a season or two. Plants build themselves out of air and water, taking only a tiny fraction of their mass from the soil.

Forests cover about a third of the land, compared with about half ten thousand years ago. (They have grown back some over the last half century in the United States, mostly because marginal farms were abandoned and trees reclaimed the land.) Like the ocean, land plants hold about three times as much carbon as the atmosphere. While oceans take many centuries to exchange this mass with the air, flora take only a few years.

As tropical societies clear the rain forest, the temperate nations have actually been growing more trees, slightly offsetting this effect. In the U.S. we have lost about a quarter of our forest cover since Columbus, and replanting occurs mostly in the South, where pine trees are a big cash crop for the paper industry. But globally we destroy a forested acre every second. Just staying even with this loss demands a considerable planting program.

Trees soak up carbon fastest when young. Planting fast-growing species will give a big early effect, but what happens when they mature? Eventually they either die and rot on the ground, returning nutrients to the soil, or we burn them. If this burning replaces oil or coal burning, fine and good. Even felling all the trees still leaves some carbon stored longer as roots and lumber.

About half the U.S. CO_2 emissions could be captured if we grew tree crops on economically marginal croplands and pasture. More forests would enhance biodiversity, wildlife, and water quality (forests are natural filters), make for better recreation, and give us more natural wood products.

Even better, one can do the cheapest part first, with land nobody uses now. This would cost about five billion dollars a year. A feel-good campaign would sell easily, with merchants able to proclaim their ecovirtue ("Buy a car, plant a grove of trees").

In the short run, this would probably work well. But trees take water, and land is limited, so this is a solution with a clear horizon of about forty years. Soaking up the world's present CO_2 increase would take up an Australia-sized land area, that is, a continent. But most such land is in private hands, so the job cannot be done by government fiat in its own territories. Still, a regional effort could make a perceptible dent in overall carbon dioxide levels.

THE GERITOL SOLUTION

The oceans comprise the other great sink of greenhouse gases; some estimates say they absorb forty percent of the fossil fuel emissions.

In coastal waters rich in runoff, plankton can swarm densely, a hundred thousand in a drop of water. They color the sea brown and green where big rivers form deltas, or cities dump their sewage. Tiny yet hugely important, plankton govern how well the sea harvests the sun's bounty, and so are the foundation of its food chain. Far offshore, the sea returns to its plankton-starved blue, a wet desert.

The oceans are huge drivers in the environmental equations, because within them the plankton process vast stores of gases. In ice ages, air CO_2 levels dropped thirty percent.

Could we do this today? Driving CO_2 down should lower temperatures, certainly. But how?

The answer may lie not in the warm tropics, but in the polar oceans. There, huge reserves of key ingredients for plant growth, nitrates and phosphates, drift unused. The problem is not weak light or bitter cold, but lack of iron. Electrons move readily in its presence, playing a leading role in trapping sunlight.

A radical fix, then, would be to seed these oceans with dissolved iron dust. This may have been the trigger that caused the big CO_2 drop in the ice ages: the continents dried, so more dust blew into the oceans, carrying iron and stimulating the plankton to absorb CO_2. Mother Nature can be subtle.

Still, such an idea crosses the momentous boundary between natural mitigation and artificial means. Here is the nub of it, the conceptual chasm. With a boast that may cost his cause dearly, the inventor of the idea, Dr. John Martin at the Moss Landing Marine Laboratories in California, said, "Give me half a tanker full of iron and I'll give you another ice age."

The captured carbon gets tied up in a "standing crop" of plankton; basically, this is ocean forestation. The CO_2 slowly dissolves into the lower waters, perhaps eventually depositing on the seabed. If we decide to stop the process, the standing crop will die off within a week, providing a quick correction.

First proposed in 1988, this "Geritol solution" has had a rocky history. Many derided it automatically as foolish, arrogant, and politically risky. But in 1996 the idea finally got tested, and performed well.

Near the Galapagos Islands lies a fairly biologically barren area. Over twenty-eight square miles of blue sea, scientists poured in 990 pounds of iron throughout a week of testing. Immediately the waters bloomed with tiny phytoplankton, finally covering two hundred square miles, suddenly green. Plankton production peaked nine days after the experiment started. A thousand pounds of iron dust stimulated over two thousand times its own weight in plant growth, an incredibly higher yield than any fertilizer ever makes on land. The plankton soaked up CO_2, reducing its concentration in nearby sea water by fifteen percent. This deficiency it quickly made up by drawing carbon dioxide from the air.

But there is some evidence that little of the newly fixed carbon actually sank. It seems to have come back into chemical equilibrium with the air. Controversy surrounds this essential point; clearly, here is where more fieldwork could tell us much.

Projections show that since this process would affect only about sixteen percent of the ocean area, probably a full-bore campaign to dump megatons of iron into the polar oceans would suck somewhere between six and twenty-one percent of the CO_2 from the atmosphere, with most recent estimates fixing around ten percent. This would dent the greenhouse problem, but not solve it entirely.

There will surely be side effects. One projected already is that fish will become more abundant, since there will be more food all the way up the chain of life. Perhaps fishing the iron-rich waters could yield enough harvest to fund the program, adding to the protein diet of many in the tropical nations. Not all side effects need be bad.

Even such partial solutions attract firm opponents. Geoengineering carries the strong scent of hubris.

Ecovirtue reared its head immediately after the 1988 proposals, well before any experiments. Many scientists and ecologists saw in it an incentive for polluters, on the Puritan model that any deviation from abstinence is itself a further indulgence. Some retaliated; Russell Seitz of Harvard said the experimenters were downplaying their results out of fear of seeming politically incorrect. "If this approach proves to be environmentally benign," Seitz said, "it would appear to be highly economic relative to a Luddite program of declaring war against fire globally."

Of course there are big uncertainties. How would the iron affect the deeper ecosystems, of which we know little? Will the carbon truly end up on the seabed? Can the polar oceans carry the absorbed carbon away fast enough to not block the process? Would the added plankton stimulate fish and whale numbers in the great Antarctic ocean? Or would some side effect damage the entire food pyramid?

Costs are easy to estimate; there is nothing very high-tech about dumping iron. Martin estimated that the job would take about half a million tons per year. Depending on what sort of iron proves best at prodding plankton, the iron costs range between ten million to a billion dollars. Total operations costs—fifteen ships steaming across the polar oceans all year long, dumping iron dust in lanes— bring the total to around ten billion dollars. This would soak up about a third of our global fossil-fuel generated CO_2 emission each year.

In 1998 an enterprising American, Michael Markels, carried out tests of iron-seeding to both lock up carbon and to pay for itself by increasing fish yields. He arranged with the Marshall Islands to at- tempt to stimulate fish production in a 100,000-square-mile area. Some estimates predict that a single kilogram of iron could yield a crop of two hundred kilograms of fish, a profitable yield for tropical nations where protein is scarce. Blanketing such vast areas with oxygen-depleting blooms is bound to cause major effects, perhaps not all to the good. Still, without experiments we will learn little.

Probably this food cycle will also produce tiny aerosols, which in turn stimulate cloud condensation. Over the long lanes left by the steaming iron spreaders, cottony clouds will form.

Is this bad? After all, clouds reflect sunlight, lessening the overall heating problem. This points to the other grand greenhouse strat- egy—altering the reflection of the planet itself.

REFLECTING ON REFLECTIVITY

What could be more intuitively appealing than simply reflecting more sunlight back into space, before it can be emitted in heat radiation and then absorbed by CO_2?

People can understand this readily enough; black T-shirts are warmer in the sun than white ones. We already know that simply by painting buildings white we keep them cooler. We could com- pensate for the effect of all greenhouse gas emission since the In- dustrial Revolution by reflecting less than one percent more of the sunlight.

Astronomers call a planet's net reflectivity the albedo, and a mere half of one percent decrease in Earth's albedo would solve the greenhouse problem completely. The big problem is the oceans, which comprise about seventy percent of our surface area and absorb more because they are darker than land.

The most environmentally benign proposal for doing this is very high-tech: an orbiting white screen, about two thousand kilometers on a side. Even broken up into small pieces, putting such parasols up would cost about $120 billion, a bit steep. As well, we would have to pay considerably to take them down if they caused some undesirable side effect. In fact, one is certain—a night sky permanently light-polluted, making astronomers and moonstruck lovers irritable.

Using more innocuous orbiting dust to reflect sunlight does not work; it drifts away, driven off by the sun's light pressure. These ideas were no doubt amusing for the National Academy panel to toy with (even scientists need distraction), but their comic aspect detracts from the point. We do not need to go into space to fix the thermostat.

Still, attention first turned to reflectors at high altitudes because much sunlight gets absorbed in the atmosphere on its way to us. Spreading dust in the stratosphere appears workable because at those heights tiny particles stay aloft for several years. This is why volcanoes spewing dust affect weather strongly.

Even better than dust are microscopic droplets of sulfuric acid, which reflect light well. Sulfate aerosols can also raise the number of droplets that make clouds condense, further increasing overall reflectivity. This could then be a local cooling, easier to monitor than CO_2's global warming. We could perform such small, controllable experiments now.

The amount of droplets or dust needed is a hundred times smaller than the amount already blown into the atmosphere by natural processes, so we would not be venturing big dislocations. And we would get some spectacular sunsets in the bargain.

The cheapest way of delivering dust to the stratosphere is to shoot it up, not fly it. Big naval guns fired straight up can put a one-ton shell twenty kilometers high, where it would explode and spread the dust. This costs only a hundredth of the space parasol idea. Rockets, balloons, and aircraft all perform worse.

But why stick to dust when we already add a perfectly good reflecting medium to the upper atmosphere as part of everyday flying—aircraft exhausts. Changing the fuel mixture in a jet engine to burn rich can leave a ribbon of fog behind for up to three months, though as it spreads it becomes invisible to the eye.

Since fuel costs about fifteen percent of airlines' cash operating expenses; for $7 million this method would offset the 1990 U.S. greenhouse emissions, quite a cheap choice. Even hiring air freight companies to carry dust and dump it high up would cost only ten times as much as this, still a small sum. Perhaps an added asset is that quietly running rich on airline fuel will attract little notice and is hard to muster a media-saturated demonstration against.

But there are, as always, side effects. Dust or sulfuric acid would heat the stratosphere, too, with unknown impact. Some suspect that the ozone layer could be affected. If a widespread experiment shows this, we could turn off the effect within roughly a year as the dust settles down and gets rained out.

Stranger ideas have been advanced. For example, making a very high altitude screen of many aluminized, hydrogen-filled balloons far above air traffic could work, and self-distribute itself over the entire globe. But the balloon parasols would cost twenty times the jet-plume approach. Ruptured balloons falling in the back garden could irritate all humanity. Imagine late night comedians using them as props. . . .

These ideas envision doing what natural clouds do already, as the major players in the total albedo picture. Just a four percent increase in stratocumulus over the oceans would offset global CO_2 emission. Land reflects sunlight much better than the wine-dark seas, so putting clouds far out from land, and preferably in the tropics, gets the greatest leverage.

Still, the most recent research shows that global averages are misleading, because climate dynamics depends on how spatially patchy reflectivity is. Even as we plan, we must keep in mind our ignorance of the complexities.

Making clouds is an old but still controversial craft. Clouds condense around microscopic nuclei, often the kind of sulfuric acid droplets the geoengineers want to spread in the stratosphere. The oceans make such droplets as sea algae decays, and the natural production rate sets the limit on how many clouds form over the seas. Clouds cover about thirty-one percent of our globe already, so a four percent increase is not going to noticeably ruin anybody's day.

Tinkering with such a mammoth natural process seems daunting, but in fact about four hundred medium-sized coal-fired power plants give off enough sulfur in a year to do the job for the whole Earth. (This in itself suggests just how much we are already perturbing the planet.)

The trouble is that coal plants sit on land, and the clouds must be at sea. A savvy international strategy leaps to mind: subsidize electricity-dependent industry on isolated Pacific islands and ship them the messiest sulfur-rich coal. Their plumes would stretch far downwind and the manufactured goods could revitalize the tropical ocean states, paying them for being global good neighbors.

A more boring approach, worked out by the National Academy panel, envisions a fleet of coal-burning ships that heap sulfur directly into the furnaces. They spew great ribbons of sulfur vapor far out at sea, where nobody can complain, and cloud corridors form obediently behind.

Best would be to use these sulfur clouds to augment at the edges of existing overcast regions, swelling them and increasing the lifetime of the natural clouds. The continuously burning sulfur freighters would follow weather patterns, guided by weather satellite data.

At first these should operate as regional experiments, to work out a good model of how the ocean-cloud system responds. Moving from science to true geoengineering could take a decade or two. This low-tech method would cost about $2 billion per year, including amortizing the ships.

The biggest political risk here lies with shifts in the weather. The entire campaign would increase the sulfur droplet content in our air by about a quarter. Probably this would cause no significant trouble, with most of the sulfur raining out into the oceans, which have enormous buffering capacity. Keeping the freighters a week's sailing distance from land would probably save us from scare headlines about sudden acid rains on farmers' heads, since about thirty percent of the sulfur should rain out each day.

Maybe some collaboration would work here. Freighters burning sulfur could also spread iron dust, combining the approaches, with some economies. Further scrutiny will probably turn up further savings; these calculations are back-of-the-envelope.

As well, the freighters would operate far from people's everyday lives, avoiding Not in My Backyard movements. This is definitely not so for, say, firing dust into the stratosphere, with the booming of naval guns.

ALBEDO CHIC

The National Academy panel found that "one of the surprises of this analysis is the relatively low cost" of implementing some significant geoengineering. Even if their rough estimates are wrong by a factor of ten, they are striking. It might take only a few billion dollars to mitigate the U.S. emission of CO_2. Compared with stopping people in China from burning coal, this is nothing.

We should not hold the 1992 panel report, thick with footnotes and layers of qualifiers, to be a road map to a blissful future. Their estimates are simple, linear, and made with poorly known parameters. They also ignore many secondary effects.

For example, forests promote clouds above them, since the water vapor they exhale condenses quickly. Growing trees to sop up CO_2 then also increases albedo, a positive feedback bonus. Is this the end of the chain? Unlikely.

Perhaps the greatest unknown is social. How will the politically aware public (those who vote, anyway) react?

If they are painted early and often as Doctor Strangeloves of the air, the geoengineers will fail. Properly portrayed as allies of science, they could become heroes.

Here a crucial factor is whether the agenda looks like another top-down contrivance, orders from the elite. A Draconian policing of illegal fuel burning will indeed look this way, but mitigation does not have to. It will play out far from people's lives, out at sea or high in the air.

Better, perhaps widespread acceptance of mitigation strategies could lead to an albedo chic, with ostentatious flaunting of white roofs, cars, the return of the ice cream suit in fashion circles. White could be appropriate after Labor Day again.

More seriously, simply adding sand or glass to ordinary asphalt ("glassphalt") doubles its albedo. This is one mitigation measure everyone could see—a clean, passive way to Do Something. Cooler roads lessen tire erosion, too.

The urban heat island effect, which drives up air-conditioning energy consumption in summer, would ease. A 1997 study by the Department of Energy showed that Los Angeles is five degrees Fahrenheit warmer than surrounding areas, mostly due to its dark roofs and asphalt. Cars and power plants add a bit, but not much;

at high noon the sun delivers to each square mile the power equivalent of a billion-watt electrical plant.

White roofs, light-colored concrete, and about $10 billion in new shade trees could cool the city to below the countryside, cutting air-conditioning costs by eighteen percent and quickly paying for itself.

About one percent of the U.S. is covered by human constructions, mostly paving, suggesting that we may already control enough of the land to get at the job. Many small measures could add to a global change. Every little bit would indeed help. This is crucial: mitigation wears the white hat. It asks simple, clear measures of everyone, before going to larger scale interventions. Grassroots involvement should be integral from the very beginning.

This should go apace with efforts at the nation-state level, especially since mitigation intertwines deeply with diplomacy. Here appearances are even more critical, given the levels of animosity between the profligate burners (especially the U.S.) and the tropical world.

Plausible solutions should stay within the National Academy of Science panel's sober guidelines. Learning more is the crucial first step, of course. This is not just the usual academic call for more funded research; nobody wants to try global experiments on a wing and a prayer.

Beyond more studies and reports, we must begin thinking of controlled experiments. Climate scientists have so far studied passively, much like astronomers. They have a bias toward this mode, especially since the discernible changes we have made in our climate have been generally pernicious. Such mental sets ebb slowly. The reek of hubris also restrains many.

But a time for limited experiments like the iron-dumping one will come. This will be the second great step as we ponder whether to become true geoengineers. Constraints must be severe to ensure clear results.

Most important, perturbations in climate must be local and reversible—and not merely to quiet environmentalist fears. Only controlled experiments, well-diagnosed, will be convincing to both sides in this debate.

Indeed, the green plume near the Galapagos Islands during the iron-spreading experiments showed this. Its larger features were best studied by satellite, which picked up the green splotch strongly against the dark blue sea. But the crucial issue of whether the carbon stayed tied up in ocean waters was poorly diagnosed. Satellites were of no help. Slightly better funding and more scientists in dispersed, small craft could have told us a lot more.

Third, careful climate modeling must closely parallel every experiment. Few doubt that our climate stands in a class by itself in terms of complexity. Though much is made of how wondrous our minds are, perhaps the most complex entity known is our biosphere, in which we are mere mayflies. Absent a remotely useful theory of complexity in systems, we must proceed cautiously.

While computer studies are notorious for revealing mostly what was sought, confirming the prejudices of their programmers, methods are improving quickly. They can explore the many side avenues of weakly perturbing geoengineering experiments. Invoking computer models as crucial watchdogs in every experiment will calm fears, at least among the elite who read beyond the headlines.

We can expect to see many technical tricks for offsetting climate change. The problem is broad and many angles of attack will emerge. For example, a principal source of atmospheric methane is bovine flatulence. An additive to their fodder might sidestep the portion of the digestive cycle that converts about ten percent of the calories into methane, both lowering food costs and eliminating the methane.

Or consider another seemingly minor fact: much of our crop production leaves behind stalks, husks, and debris rotting in the fields. Usually this has little effect upon soil improvement. Compacting these leftovers and stopping them from returning their carbon to the air, then, could offset our fossil fuel burning. Physicist Robert Metzger has suggested dropping such waste into the deep ocean sinks, where cold and lack of oxygen would effectively cut them out of the atmospheric cycle. A rough estimate shows this could compensate for a major fraction of world CO_2 emissions.

Such interventions into our business-as-usual world could be minor in effort, but potentially large in effect. Geoengineering need not be obnoxious, or even obvious. Once society accepts the principle of mitigation, ingenuity can come into play.

Still, going from the local to the global is fraught with uncertainty, certain to inspire much anxiety. We will always be uncertain stewards of the Earth. And probably the greenhouse shall not be our last problem, either. We are doing many things to our environment, with our numbers expected to reach ten billion by 2050. What new threats will emerge?

Fresh disasters shall probably spring from the many synergistic effects that we must trace through the geophysical labyrinth. Once we become caretakers, we cannot stop. The large tasks confronting humanity, especially the uplifting of the majority to some semblance of prosperity, must be carried forward in the shadow of our stewardship.

PURITANS AND PROPHETS

Americans have a deep propensity for pursuing morality into the thickets of science. The cloning of a sheep in Scotland provokes stern finger-pointing by archbishops and ethics experts alike. This we must expect as scientific topics become ever more complicated and freighted with profound implications.

So it inevitably shall be for greenhouse gases. Of *course* it would be far better to abstain from burning so much. The fossil carbon resources are a vast heritage whose utility, convenience, and endless applications we—or rather, surely our great-grandchildren—shall not see again.

But the experience of the last several decades holds no grounds for more than a sliver of optimism. Modern prosperity is built upon cheap, handy energy. Billions of new people strain the carrying capacity of the planet—which some globalists believe we have already exceeded. Surely this shall exacerbate our habit of digging energy wealth from the ground, the burnable corpses of early life. We are addicted to fossil fuels, and the ambitions of billions to come will hinge upon more use of them, not less.

In the end, the deep Puritan impulse to scold and condemn for rising fossil fuel use will very probably fall upon deaf ears in both the emerging and the advancing nations. It has not fared well among its natural constituency, those who are far more easily moved by campaigns to end smoking, save the whales, or stop sullying the city air.

But among the able nations, those who have the foresight to grasp solutions, an odd reluctance pervades the policy classes. The past shouts but the future can only whisper. Ralph Cicerone, a noted atmospheric physicist, noted that "many who envision environmental problems foresee doom and have little faith in technology, and therefore propose strong limits on industrialization, while most optimists refuse to believe that there is an environmental problem at all."

Having sinned against Mother Nature inadvertently, many are keenly reluctant to intervene knowingly. Nobel laureates have generally sided with the Puritan position, adding considerable weight to the cause of abstention.

At root, we see ourselves as the problem; only by behaving humbly, living lightly upon our Earth, can we atone. Here most scientists and theologians agree, at least for now. Alas, such a future world would probably be hostile to science, for already technology is routinely blamed for pollution, not human appetites. People fear and fault the new, while finding the old cozy, even if in fact, like coal burning, it is more dangerous.

Quite probably, the next century shall see a protracted battle between the prophets who would intervene and the moralists who see all grand scale human measures as tainted. In this view, any human intrusion onto so august a stage must be wrong, classic hubris. Yet again, they shall argue, our scientists and engineers will be brought forward to solve a problem that properly belongs to the moral sphere. That our greenhouse problem stems from inadvertently acting on this stage simply proves, to many, that any such action is too arrogant; thus, they preclude advertent action.

Even now many advance their argument that to even speak of geoengineering encourages the unwashed to even more excess, since the masses will think that once again science has a remedy at hand.

Some, though, will say quietly, persistently, *Well, maybe science does.* . . .

OUR WORLD AS MESSAGE

A decision to face our historical role as stewards, and intervene knowingly, globally, would be a watershed in human history.

Managing the entire planet, and so ensuring an enduring legacy, is surely a more striking prospect than the makers of any deep time message have ever contemplated. Yet doing so will make an explicit connection with distant eras, which unknowingly bequeathed us an altered landscape. Such links are made concrete in the transformed world we now inhabit, a "natural" environment far removed from that of our Pleistocene ancestors.

Once taken up, such a monumental task cannot be put down. As long as we exert so powerful a presence, we shall have to be mindful of it. Perhaps the worldview needed can emerge from such philosophers as Martin Heidegger. In his perspective, technology "enframes" us, leading humanity to more and more see the world as a *bestanden*, a "standing reserve." In turn, we regard other humans as resources, reducing our unconscious respect for both them and their environment. Heidegger is abstruse but influential, and may point the way toward a philosophical shift necessary for true stewardship.

It is small consolation to estimate that we may well consume nearly all fossil fuel reserves within five hundred years, and certainly within a millennium. This assumes we will use them as our principal energy store. Of course, other sources might turn up. Within a century or so our descendants may view with horror the burning of the long hydrocarbon chains bequeathed by geology. They may feel oil should properly be employed as lubricants rather than plundered for simple energy content.

Should world civilization fall, its hand would slip from the helm, and the planet could fare badly. But that has been so since we erected civilizations; a fall backward would sacrifice some, perhaps most, of our softened population. Such a shrinkage followed the slow ebbing of Rome. Historically, such retreats have little to recommend them.

Still, if we elect to guide the planet so that it may support so many of us in high technological civilizations, we will have to exert wisdom and cleverness throughout centuries-long agendas. Nature in her largeness and largesse responds with ponderous grace on the scale of centuries or more. Our frenetic energies expend themselves within a decade or two, the typical scale for even the most far-reaching customary projects.

This mismatch in time scales implies that we must cultivate a farsighted judgment typically unknown in our societies, ancient or modern. Becoming aware of how we leave our mark on the world will demand a cultural evolution profound and, to many, puzzling. Yet making such a transformation may well be our era's most lasting deep time monument.

AFTERWORD

. .
.

Here on the level sand
Between the sea and land,
What shall I build or write
Against the fall of night?
Tell me of runes to grave
That hold the bursting wave,
Or bastions to design,
For longer date than mine.
—A. E. HOUSMAN

A visit to a Pleistocene cave in southern France reveals the past in subtle ways. Paintings on the cave walls and ceiling show a pack of wild horses galloping along a ledge, while vivid antlered reindeer leap toward the viewer from nearby walls. Bison scratched into stone show fine-line features of nostrils, eyes, and hair. Big-bellied horses lope toward us on short legs.

These are not crude sketches. A big rocky bulge forms the muscular shoulder of a bison. A cow's body follows quite naturally a long, deep depression in one wall. Cleverly drawn animals blend, sharing a natural line in the wall. A ceiling frieze of small reindeer seem simply rendered under a flashlight's direct beam, but when the light angles away, the racks of their antlers follow the crests of slightly raised ridges in the rock.

Some prehistoric master saw the essence of these animals embedded in the chance curves of the cave. Then he called them forth to the eye, using negative space in ways we do not witness again until the work of the sixteenth century.

These signals across tens of millennia carry a heady sense of graceful intelligence. We know well enough what animals lived then, but only in such paintings can we delve into the cerebral wealth of our ancestors. Whether the artist intended them as such, these paintings, then, are the best sort of deep time messages, conveying wordless mastery and penetrating sensitivity across myriad millennia and staggeringly different cultures.

It is sobering to contemplate that our distant heirs may know us best not by our Michelangelos or Einsteins or Shakespeares, but by our waste markers, our messages aboard space craft, our signatures upon the soil and species, or our effect upon their landscape, descended from ours.

Yet that is a proposition we must entertain.

In writing this book I have made my views quite apparent, attempting to see the phenomena of deep time from many angles and not suppressing my own, perhaps quirky outlook. My personal experiences I have rendered frankly, seeking to trap my momentary glimpses into how we look down the long barrel of time's weaponry.

I hope specialists in the many areas I touch will not find my methods and summaries too rough. Wherever possible, I have cemented my intuitions with travel, visits, and detailed consultation. I feel that conclusions won from experience have a solidity that armchair ruminations cannot. I have visited many ancient sites, seeking the odd and unexpected angle on the buried past, the oblique, even accidental understanding. Only by trying varying perspectives can we grasp how our culture may someday look to others vastly different, and perhaps better experienced.

John Horgan has used "ironic science" to describe speculative discussion that cannot be tested, as proper science should, in *The End of Science*—a book that combines effective writing with mistaken logic, for science is nowhere near any foreseeable end. The book you hold is not "ironic"; rather, I should call it "improper" science. I sought here to use ideas and findings from many scientific areas—social, biological, and physical—to attempt a sort of understanding that is not scientific in the proper sense, yet is informed by it. Instead, using a scientific eye to peer through the lens of deep time, we can glimpse humanity in long perspective, across cultures and concerns that, though vital in their moving moments, are passing phenomena. Perhaps not every summary and judgment I make is even sufficiently informed, in the view of experts, a risk this kind of improper venture must take.

Improper science need not be popular, for its concerns do not necessarily parallel the common interests of its day. An example of this is J. D. Bernal's *The World, the Flesh and the Devil*, which examined our long-term prospects in terms that seemed bizarre in 1929 but resonate strongly today. It was an early expedition into the realms of deep time. This slim book found its proper and continuing audience long after its first edition went out of print.

Nor is such improper science "fiction about science" or science fiction; it is science thinking about itself as a human agenda in the dimension of time. Improper science is broad and may not need to be deep. It necessarily speculates, making ranging forays into territories seldom illuminated coherently in our era of intense narrowness.

Outright speculation is not rare in proper science, but it often arrives well disguised. Sometimes it is a short-term claim to a notion awaiting exploration. When James Watson and Francis Crick laconically noted in the last sentence of their paper reporting the discovery of DNA's double helix, they saw speculative implications for reproduction: "It has not escaped our notice that the specific pairing we have postulated immediately suggests a possible copying mechanism for the genetic material."

In Parts III and IV, I have deliberately used a long view to suggest attacks upon the largest problems we can glimpse on the horizon of the next millennium. These proposals will be much criticized; I intend them to be. Some will say they are fatalistic; they seem to me merely prudent, in light of our own nature. They are sincere attempts to suggest, by example, that increasingly our advanced societies must take a longer view. For if not us, who will?

Our time can benefit from the vistas made possible by science. When hatred and technology can slaughter millions in months or even minutes, such terrors deprive life of that quality made scarce and most precious to the modern mind: meaning. Deep time in its panoramas redeems this lack, rendering the human prospect again large and portentous. We gain stature alongside such enormities.

Though I have here deplored the Kilroy Was Here impulse, I do not suggest that Kilroy's followers were not expressing strongly felt emotions. Their gestures against the inevitable are merely futile, conveying little. Our names are surely the least aspect of our selves.

Considering our position in the long roll of epochs demands breadth transcending the momentary and the passingly personal. To reverse a famous saying of Newton's, I would hope that our grandchildren can fondly say of us, that if they have seen farther than our generation, it will be because they are standing taller.

* * *

Certain professions lend their followers an intuitive grasp of long duration.

Archeologists sense the rise and fall of civilizations by sifting through debris. They are intimately aware of how past societies mismanaged their surroundings and plunged down the slope of collapse, sometimes with startling speed.

Biologists track the extinction of whole genera, and in the random progressions of evolution feel the pace of change that looks beyond the level of mere species such as ours. Darwinism invokes cumulative changes that can act quickly on insects, while mammals take millions of decades to alter. Our own evolution has tuned our sense of probabilities to work within a narrow lifetime, blinding us to the slow sway of long biological time. This may well be why the theory of evolution came so recently; it conjures up spans beyond our intuition. On the creative scale of the great, slow, and blunt Darwinnowings such as we see in the fossil record, no human monument can sustain. But our neophyte species can now bring extinction to many, which is forever.

In their careers, astronomers discern the grand gyre of worlds. But planning, building, flying, and analyzing one mission to the outer solar system commands the better part of a professional life. Future technologies beyond the chemical rocket may change this, but there are vaster spaces beckoning beyond, which can still consume a career. A mission scientist invests the kernel of his most productive life in a single gesture toward the infinite.

Those who study stars blithely discuss stellar lifetimes encompassing billions of years. In measuring the phases of stellar mortality they employ the many examples, young and old, that hang in the sky. We see suns in snapshot, a tiny sliver of their grand and gravid lives caught in our telescopes. Cosmologists peer at distant reddened galaxies and see them as they were before Earth existed. Observers measure the microwave emission that is relic radiation from the earliest detectable signal of the universe's hot birth. Studying this energetic emergence of all that we can know surely imbues, and perhaps afflicts, astronomers with a perception of how like mayflies we are.

No human enterprise can stand well in the glare of such wild perspectives. Perhaps this is why for some, science comes freighted with coldness, a foreboding implication that we are truly tiny and insignificant on the scale of such eternities. Yet as a species we are young, and promise much. We may come to be true denizens of deep time.

Though our destiny is forever unclear, surely if we persist for another millennium or two we shall fracture into several species, as our grasp of our own genome tightens. We will dwell on the scale of a hastening evolution, then, seizing natural mechanisms and turning them to our own tasks. In this sense we will emerge as players in the drama of natural selection, as scriptwriters.

Our ancient migrations over Earth's surfaces have shaped us into "races" that cause no end of cultural trouble, and yet are trivial outcomes of local selection. Expansion into our solar system would exert selective pressure upon traits we can scarcely imagine now, adaptations to weightlessness, or lesser gravity, or other ranges of pressure or temperature. In this context, we will need long memories of what we have been, to keep a bedrock of certainty about what it means to be human. This is the work of deep time messages, as well.

The larger astronomical scale, too, will beckon before us in such a distant era, for well within a millennium we will be able to launch probes to other stars. To ascend the steps of advanced engineering and enter upon the interstellar stage will portend much, introducing human values and perceptions into the theater of suns and solar systems. The essential dilemma of being human—the contrast between the stellar near-immortalities we see in our night sky and our own all-too-soon, solitary extinctions—will be even more dramatically the stuff of everyday experience.

What changes might this presage? We could lend furious energies to the pursuit of immortality, or something approximating it. If today we eliminated all disease and degeneration, accidents alone would kill us within about fifteen hundred years. Knowing this, would people who enjoyed such lifetimes nonetheless strive for risk-free worlds, to further escape the shadow of time's erosions?

On the scale of millennia, threats and prospects alter vastly. Over a few thousand years, the odds that a large asteroid or comet will strike the Earth, obliterating civilization if not humanity, become considerable. But within the next century, as our ability to survey the solar system and intervene there grows to maturity, we will be able to protect our planet (or even others) from such risks.

This marriage of space science and planetary protection will seem inevitable by then, for it shall occur in the same era that we learn, perhaps by rudely administered punishments, to be true stewards of the planet. The impulse to do so will spring from a similar sense of the perspectives afforded, if we heed, by pondering deep time. A steward must look long.

We are ever restless, we hominids. It is difficult to see what would finally still our ambitions—neither the stars, nor our individual deaths, would ultimately form a lasting barrier. The impulse to push further, to live longer, to journey farther—and to leave messages for those who follow us, when we inevitably falter and fall—these will perhaps be our most enduring features.

Still, we know that all our gestures at immortality—as individuals or even as a lordly species—shall at best persist for centuries or, with luck, a few millennia. But ultimately they shall fail.

Intelligence may even last to see the guttering out of the last smoldering red suns, many tens of billions of years hence. It may find a way to huddle closer to the dwindling sources of warmth in a universe that now seems to be ever-expanding, and cooling as it goes. Whether intelligence can persist against this final challenge, fighting the ebb tide of creeping entropy, we do not know.

But humans will have vanished long before such a distant waning. That is our tragedy. Knowing this, still we try, in our long twilight struggles against the fall of night. That is our peculiar glory.

ACKNOWLEDGMENTS

This book spans many fields, in most of which I am not expert. I have made generalities that may reek of oversimplification to those better acquainted with an area. I beg some forbearance; capturing an essence without overburdening with detail is an art. Where possible I have enlisted help and advice from many who know more, and I thank them for their expertise. Any errors are of my doing, not theirs.

I am indebted to those who contributed valuable comments and suggestions: Mark Martin, Martin Pasqualetti, Kate Trauth, Craig Kirkwood, Michael Brill, Harry Otway, John Paul Brusky, Freeman Dyson, Marvin Minsky, Oliver Morton, Martyn J. Fogg, Jeffrey A. McNeely, Steve Harris, Michael Rose, David Brin, Oliver Ryder, Kurt Benirschke, Harold Morowitz, Murray Gell-Mann, Gregory Fahy, Hugh Hixon, Michael Darwin, Robert Stone, E. O. Wilson, Christopher Wills, Martin Rees, Virginia Trimble, David Givens, Michael Soulé, Ralph Cicerone, Virginia Postrel, John Casti, Hal Lewis, Elisabeth Malartre, Marina Brown, Marvin Johnson, Bruce Murray, Ken Forman, Steve Postrel, Paul Levinson, Tom Shippey, James Benford, Mark Benford, Louis Friedman, Charles Kohlhase, Tobias Owen, Carolyn Porco, Louis Narens, and Michael Cassutt.

I am especially grateful to Jon Lomberg for many discussions, insights, and much help. (His own views on these matters will appear in a forthcoming book.) My wife Joan suffered patiently through many drafts. My editor, Jennifer Brehl, made many important suggestions, crucial to my first work of nonfiction.

Gregory Benford
May 1998

REFERENCES

INTRODUCTION

Dean MacCannell notes . . . in *The Tourist: A New Theory of the Leisure Class*, Schocken Books, 1976, p. 3.

A puzzle of far antiquity . . . see Jeffrey D. Kooistra, *Analog*, January 1998. Monument construction followed swiftly upon the invention of agriculture, roughly ten thousand years ago, though for evidence of earlier origins five thousand years before in Asia consult "An Earlier Agricultural Revolution," Wilhelm G. Solheim III, *Scientific American*, April 1972. By way of contrast, the earliest exactly dated year in history is 4241 B.C.

a citizen of the year 5000 B.C. . . . Herein I use the customary scheme of B.C. (Before Christ) and A.D. (Anno Domini), mostly because it *is* customary. There are other dating methods used by Islam, China, and the Jews (Anno Mundi), among others, but none are now widespread. Such absolute dating systems are useful over deep time scales. Relative dating, as accomplished by radioactive decay measurements, must carry a tag in absolute units showing when the dating was done. Current museum fashion for replacing B.C. with B.C.E. for Before Common Era (with C.E. for A.D.) seems pointless, though to some its goal of finding a dating scheme that does not favor a particular culture appears worthy. When asked what is common about our Common Era, it turns out to be dating from the birth of Christ, so we are back where we started. More seriously, navigating deep time places a premium on predictability. B.C. and A.D. are now time-honored, in use about a millennium. Introducing a new, suspiciously similar scheme promises to simply confuse future historians and archeologists. It seems likely, given the passing political nature of the change, that more methods will crop up to accomplish similar short-term goals; they all should be deplored.

Individual recollections of the past . . . see Paul Fussell's *The Great War in Modern Memory*, London: Oxford University Press, 1975.

Astronomical time itself . . . see *The Bias of Communication* by Harold A. Innis, University of Toronto Press, 1951, p. 62. [For aspects of geological time, Stephen Jay Gould's *Time's Arrow, Time's Cycle* (Harvard University Press, 1987) is an engaging overview. The term "deep time" was first used in John McPhee's *Basin and Range* (Farrar, Straus, and Giroux, 1980) to describe the depths of geological time.]

Social time . . . see P. A. Sorokin and R. K. Merton, *American Journal of Sociology*, XLII, p.1936.

Chinese concepts . . . see R. K. Merton in *Twentieth Century Sociology*, ed. by G. Gurvich, 1945, p. 387.

media reflect emphasis on time or space . . . see *Empire and Communications* by Harold A. Innis, University of Toronto Press, 1972, p. 7. Historically, long distance travel tends to break up cyclical (agriculture-centered) time, with its vertical sense of the heavens as the eternal clockwork. It replaces them with a linear time sense and horizontal measures; see also *Topophilia* by Yi-Fu Tuan, Prentice-Hall, 1974.

technologies are attempts to contest the ordinations of time . . . The sociologist Paul Levinson has observed this in *Mind At Large*, JAI Press, 1988, p. 138.

folk memory is surprisingly long-lived . . . as discussed in Jared Diamond's admirable *Guns, Germs and Steel*, Norton, 1997.

Nietzsche disdained . . . *The Use and Abuse of History,* F. Nietzsche, trans. Adrian Collins, 1985, pp. 14–17.

Lewis Mumford pronounced "monumentalism" dead . . . in Mumford's *The Culture of Cities*, Harcourt Brace, New York, 1938, p. 434.

some find memorials destructive . . . see *The Texture of Memory* by James E. Young, Yale University Press, 1993, Introduction. Also, Pierre Nora's "Between Memory and History," in *Representations* 26 (1989).

Thomas Campbell's "The River of Life" . . . in *Immortal Poems of the English Language*, Oscar Williams, ed., Simon & Schuster, 1952. While poetry frames and evokes our experience in politely moral terms, there is also evidence that our sense of accelerating time may come from diminishing dopamine production, which gradually resets our internal time clock as we age; see *Lancet*, Feb. 7, 1998.

William Faulkner . . . The quotation is from an interview in *Writers at Work*, First Series, ed. Malcolm Cowley, 1958.

the cave markings . . . I drew upon observations in *How to Deep Freeze a Mammoth* by Bjorn Kurten, Columbia University Press, 1992. A good guide is *Journey Through the Ice Age* (University of California Press, 1998) by Paul G. Bahn and Jean Vertut, an excellent overview of cave art's possible meanings, illustrated by many fine examples.

the Seven Wonders . . . An enjoyable tour of how ancient structures were built is L. Sprague de Camp's *The Ancient Engineers*, Dorset Press, 1963. *Then and Now* by Stefania Perring and Dominic Perring (Macmillan, 1991) shows great monuments of antiquity, then overlays them with historically informed views of how they probably looked when fresh. The sense of loss so engendered is palpable. See also *Roman Imperial Architecture* by J. B. Ward-Perkins, Penguin, 1981. A popular, rhapsodic introduction to the Seven Wonders is Leonard Cottrell's *Wonders of Antiquity*, Longmans, 1960.

We read messages everywhere . . . for background, see *Why Buildings Stand Up* (1980) and *Why Buildings Fall Down* (1987) by Mario Salvadori, Norton. On architecture and transcendance consult p. 168 and elsewhere in *Topophilia* by Yi-Fu Tuan, Prentice-Hall, 1974.

Somehow this ring of stones . . . A useful guide is John North's *Stonehenge: Neolithic Man and the Cosmos*, 1996, HarperCollins. My remarks draw also upon a review by Tom Shippey, *London Review of Books*, 12 December 1996.

The oldest reliably dated structure in North America . . . see *Science* 277, p. 1761 (1997), and the Los Angeles *Times*, Sept. 19, 1997, p. A25.

. . . a man covered by ice for 5,300 years . . . Konrad Spindler's *The Man in the Ice* (Harmony Books, 1994), gives a thorough study, including an appendix on permafrost mummies.

a striking depiction of Alexander's father . . . in *Making Faces: Using Forensic and Archeological Evidence* by John Prag and Richard Neave, Texas A&M University Press, 1997.

Time capsules embody a modern faith . . . See *Newsweek*, April 7, 1997, p. 10, and *Harper's*, April 1995, p. 30. Oglethorpe University's International Time Capsule Society (4484 Peachtree Road, N.E., Atlanta, GA 30319), issues "most wanted" lists of lost capsules and solicits information about them.

. . . an article in *Scientific American* in 1936 . . . titled "Today-Tomorrow," November.

The Long Now Foundation . . . consult www.longnow.org

Media and their messages fade from our world . . . Consult the Dead Media material shepherded by Bruce Sterling at http://www.mediahistory.com/dead/archive.html

Dyson . . . from *Imagined Worlds* by Freeman Dyson, Harvard University Press, 1997.

These scales are handily written in powers of ten . . . Nigel Calder's *Time-scale* (Viking, 1983), provides an engrossing look at the entire known timescale.

• PART ONE

Envisioning Tomorrow . . . Our scenarios appear in Sandia Laboratory's publication, *Expert Judgment on Inadvertent Human Intrusion into the Waste Isolation Pilot Plant*, Sandia Report SAND90-3062, 1993. A later assessment is *Effectiveness of Passive Institutional Controls in Reducing Inadvertent Human Intrusion into the Waste Isolation Pilot Plant for Use in Performance Assessments*, WIPP/CAO-96-3168, 1996.

The Marker panel reported in Sandia's *Expert Judgment on Markers to Deter Inadvertent Human Intrusion into the Waste Isolation Pilot Plant*, Sandia Report SAND92-1382, 1993. These cite many useful references, including earlier government reports. Many of my points I have drawn from the drafts of the various teams of this panel, and discussions with them.

Many of my observations on lessons learned from archeology stem from David B. Givens's "From Here to Eternity: Communicating with the Distant Future," *Et Cetera* 39, pp. 159-79, and "Using the Past to Protect the Future," by Maureen F. Kaplan and Mel Adams, *Archeology* , Sept. 1986, pp. 51-54. Both these articles come from early deliberations on the Pilot Project. On Newgrange and similar cases, see B. Fagan, *Archeology* 47, No. 5, pp. 16-17, 1994. On astronomy and the ancients, see E. C. Krupp's *Echoes of the Ancient Skies: The Astronomy of Lost Civilizations*, Harper & Row, 1983, and references therein. G. S. Hawkins's 1973 *Beyond Stonehenge* (Harper and Row) is a reliable guide.

As physicist Bernard Cohen discussed . . . See his paper, "Discounting in Assessment of Future Radiation Effects," in *Health Physics* 45, 687 (1983). Also, his *The Nuclear Energy Option: An Alternative for the '90s* (Plenum, 1990), discusses many of the technical and risk issues behind the Pilot Project. *Physics Today* devoted a special issue to the waste problem, June 1997.

Hal Lewis has remarked . . . in his thoroughgoing *Technological Risk*, Norton, 1990, p. 33.

. . . **many physicists who scoffed** . . . Decision-making with quantitative knowledge, Hal Lewis discusses in *Why Flip a Coin?*, John Wiley, 1997; see especially Chapter 13.

. . . **two separate second panels to discuss the marker problem** . . . Team A members were Dieter G. Ast, Michael Brill, Ward Goodenough, Maureen Kaplan, Frederick Newmeyer, and Woodruff Sullivan. Team B members were Victor Baker, Frank Drake, Ben Finney, David Givens, Jon Lomberg, Louis Narens, and Wendell Williams. Both teams had physical scientists, anthropologists, artists, and social scientists.

The Department of Energy certified the Waste Isolation Pilot Project (WIPP) for use in May 1998. Westinghouse corporation will provide final designs for markers, to be reviewed as planning proceeds for eventually erecting markers in 30 to 40 years.

• PART TWO

The Voyager Interstellar Record . . . is described in detail in *Murmurs of Earth* by Carl Sagan, Ann Druyan, Timothy Ferris, Jon Lomberg, and Linda Salzman Sagan, Random House, 1978. The book and a copy of the record are available from the Planetary Society, Pasadena, CA 91106.

The disk opens . . . A description appears in "Visions of Mars: A Cultural and Scientific Experiment on Mars '96," Louis Friedman, Walker Giberson, Jon Lomberg, and Terry Cole, Proceedings of the 45th Congress of the International Astronomical Federation, 1994.

"The Visual Presentation of Science" . . . appears in *Carl Sagan's Universe*, a memorial volume published by Cambridge University Press, 1997, with material relevant to this book.

Tor Norretranders calls this *exformation* . . . in *The User Illusion*, Viking, 1998.

Brains often must decipher the visual world . . . A concise overview is *Brainscapes* by Richard Restak, Hyperion, 1995. His earlier books go further.

Guillermo Lemarchand and Jon Lomberg . . . Their paper appears in *Astronomical and Biochemical Origins and the Search for Life in the Universe*, ed. C. B. Cosmovici, S. Bowyer, and D. Wertheimer, 1998.

. . . **this Golden Section** . . . For examples of mathematical forms in nature see *The Seven Mysteries of Life* by Guy Murchie, Houghton Mifflin, 1978.

New Scientist . . . in Volume *145*, 18 February 1995, p. 7.

the odd creatures in our fossil record . . . see Stephen J. Gould's *Wonderful Life: the Burgess Shale and the Nature of History*, Norton, 1989.

communicating with bright chimpanzees . . . Roger Fouts's claims are well detailed in his *Next of Kin* (Morrow, 1997) and challenged in work on bonobo chimpanzees in *Kanzi: The Ape at the Brink of the Human Mind* by Sue Savage-Rumbaugh and Roger Lewin, John Wiley, 1994. *Animal Minds* by Don Griffin (University of Chicago Press, 1992) makes a strong argument for discernible continuity between us and many other animals.

Objects, Causes, and Goals. . . . This argument Minsky sets forth in his classic *The Society of Mind*, Simon & Schuster, 1986.

cognitive universals . . . See Narens's "Surmising Cognitive Universals for Extraterrestrial Intelligences," in *Astronomical and Biochemical Origins and the Search for Life in the Universe*, ed. C.B. Cosmovici, S. Bowyer, and D. Wertheimer, 1997. Artificial realms like mathematics and theology are built from the start to be devoid of interesting inconsistency. This necessarily means that their cool abstractions will not convey much of human experience.

LEGEOS satellite . . . is described in *Murmurs of Earth*, 1978, Random House.

a French artist announced plans . . . *Nature* 391, 1998, p. 112.

Comedy and Tragedy . . . All quotations given are either documented or were transcribed from conversation at the time by myself. Where individuals did not wish to be identified with their quotations, I used only general descriptions of their role.

• PART THREE

. . . the work of E. O. Wilson and others . . . see *The Diversity of Life* by E. O. Wilson, Harvard University Press, 1992; W. V. Reid and K. R. Miller, *Keeping Options Alive: The Scientific Basis for Conserving Diversity*, World Resources Institute, Washington, D.C., 1989. Also, N. Myers, *Deforestation Rates in Tropical Forests and Their Climatic Implications*, Friends of the Earth, London, 1989. A projection appeared in "The Future of Biodiversity," Stuart Pimm, Gareth Russell, John Gittleman, and Thomas Brooks, *Science* 269, p. 347. A sobering overview is Peter Ward's *The End of Evolution*, Bantam, 1994. No less serious but not solemn is *Last Chance to See* by Douglas Adams, Harmony, 1990.

There are far too few taxonomists . . . see Paul Ehrlich and E. O. Wilson, *Science* 253, 1991, pp. 758-62. See also pp. 736-38.

. . . saving traditional domesticated varieties . . . see Michael Soule, *Science* 253, 1991, pp. 744-50.

Worldwide, one in every eight plant species is threatened . . . from an analysis contained in the World Conservation Union's "Red List" of 1998, which took twenty years to assemble. See also *Protection of Global Biodiversity*, L.D. Guruswamy and Jeffrey A. McNeely, eds., Duke University Press, 1998.

Sampling of tropical trees by insecticidal fogs . . . is common and dangerous to the biologists; see *Life Above the Jungle Floor* by Donald Perry, Firesign, 1986, p. 144. A good introduction to tropical rain forests is *Tropical Nature* by Adrian Forsyth and Ken Miyata, Scribner's, 1984; also references therein.

we prefer direct in situ preservation . . . though recent collections have not fared well; see Peter Raven, "Research in Botanical Gardens," *Botanische Jahrbuecher* 102, pp. 52-72.

zoos are unhappy solutions . . . see *The Modern Ark*, by Vicki Croke (Scribner's, 1997), for telling criticism of zoo limitations.

"Nature's rules are universal . . ." from *The Time Before History: 5 Million Years of Human Impact* by Colin Tudge, Scribner's, 1996.

. . . temperatures below the "glass point" . . . see G. Fahy and J. Saur, *Cryobiology* 27, 1990, pp. 492-510, and references therein.

The problem of recovering cells from frozen samples . . . for a quick sketch of work done, see *Scientific American*, Oct. 1989, p. 37.

We have been stewards . . . *Nature's Keepers: The New Science of Nature Management* by Stephen Budiansky (Free Press, 1995) argues this case.

Freeman Dyson has suggested . . . in private correspondence.

Already a crash program . . . see L. Cavalli-Storza, Cold Spring Harbor Symp., Quant. Biol. 51, 1986, pp. 411-417.

Such records may allow . . . see J. M. Diamond, *Nature* 352, 1991, p. 567.

closed populations (the "founder effect") can be mapped and managed . . . see Oliver Ryder, *Nature* 331, 1988, p. 396.

Paul Ehrlich suggested . . . in *Systematic Biology* 13, pp. 109-23.

The "frozen zoo" of San Diego . . . see Kurt Benirschke, *Zoo Biology* 3, 1984, pp. 325-28.

Cryonic mouse embryo banking . . . see P. H. Glenister, D. G. Whittingham, & M. J. Wood. (1990) *Genetic Research Cambridge* 56, pp. 253-258.

. . . the criteria of Vernon Heywood . . . see his paper in *Conservation of Plant Genes*, 1992, pp. 1-14.

John Terborgh illustrates . . . in *Diversity and the Tropical Rain Forest*, Princeton University Press, 1992, p. 15.

Terry Erwin to his 1988 estimate . . . in *The Tropical Forest Canopy*, National Academy Press, p. 127, 1988.

Harold Morowitz has remarked . . . in *Science* 253, 1991, pp. 752-54.

Worldwide there are more than seven hundred documented seed collections . . . discussed in "Seed Banks and Molecular Maps: Unlocking Genetic Potential from the Wild," by Steven D. Tanksley and Susan R. McCouch, *Science* 277, 22 August 1997, p. 1063. These fall prey to local disasters, however.

. . . collections themselves are threatened . . . *Science* 273, 27 September 1996. See also an account of how the Yale University collection of insect-borne viruses has gone begging for a home, in *Science* 278, p. 1878. While few will weep for lost viruses, in fact they are essential in health studies, yet Yale found little support for retaining them.

A unique all-taxa survey of Costa Rica recently collapsed . . . *Science* 276, p. 893. A Smoky Mountains survey may come about; *Science* 278, p. 1871.

Tropical Tactics . . . Also, smaller sections left after general clearing do not survive well, or support many species; see *Science* 278, 7 November 1997, p. 1016.

once published by the National Academy . . . in the *Proceedings of the National Academy of Sciences USA* 89, 1992, pp. 11098-11101.

attracted endorsements . . . editorial in *The Economist* 327, no. 7806, p. 17, and *Scientific American*, "Noah's Freezer," March 1993.

Mark Martin of Occidental College discussed . . . as he reported in T.R.E.E. *10*, p. 227.

• PART FOUR

Portions of this Part appeared in different form in *Reason*, November, 1997.

Virginia wasn't virgin . . . Stephen Budiansky treats this in *Nature's Keepers*, Free Press, 1995, Chapter 5.

Slash-and-burn farming already thrived . . . see *New York Times*, July 27, 1993. For a fuller discussion, *Vestal Fire* by Stephen J. Pyne, University of Washington Press, 1997, and his earlier *World Fire*, traces how systematic burning makes confusingly freighted terms like "pristine" and "virgin forest" demand historical inspection.

Archeologists know . . . see *The Time Before History: 5 Million Years of Human Impact* by Colin Tudge, Scribner's, 1996.

Ubar suffered a literal collapse . . . described in *The Road to Ubar* by Nicholas Clapp, Houghton Mifflin, 1998.

Daniel Botkin remarks, surveying our true natural history . . . in his *Discordant Harmonies: A New Ecology for the Twenty-first Century*, Oxford University Press, New York, 1990, pp. 52, 177, 191, 193. To include cultivation in a "natural" aesthetic, see Paul Shepard's *The Cultivated Wilderness: Or, What Is Landscape?*, MIT Press, 1997, p. 233.

Projections show China increasing its burning . . . "Who Will Fuel China?" by Thomas E. Drennen and Jon D. Erickson in *Science* 279, 1998, p. 1483. Largely, China itself, so no other nations can constrain its aims.

Climate experts doubt . . . *New York Times*, Nov. 3, 1997, "Experts Doubt a Greenhouse Gas Can Be Curbed."

Such a triggered shift . . . see William Calvin, *Atlantic*, January 1998.

climatic dirty tricks . . . as detailed in "Environmental Warfare," a white paper of the Westing Associates, 1996.

A little-noticed 1992 National Academy of Sciences panel . . . published its extensive calculations in *Policy Implications of Greenhouse Warming: Mitigation, Adaptation, and the Science Base*, National Academy Press, Washington, D.C., 1992. This 918-page treatment still only scratches the surface. See also for geoengineering the definitive text, *Terraforming* by Martyn J. Fogg, Society of Automotive Engineers, Warrendale, PA, 1995.

In 1998 an enterprising American . . . see *Scientific American*, April 1998, p. 33.

A 1997 study by the Department of Energy showed . . . "Painting the Town White—and Green," by Arthur Rosenfeld, Joseph Romm, Hashem Akban, and Alan Lloyd in *Technology Review*, MIT, February 1997.

Physicist Robert Metzger has suggested . . . in personal correspondence.

an odd reluctance pervades the policy classes. . . . Intervention often provokes automatic reactions, as in the letters to *Nature* following the short paper proposing that mitigation might work, by Ralph Cicerone, Scott Elliott, and Richard Turco, *Nature* 356, 1992, p. 472.

such philosophers as Martin Heidegger . . . See the title essay in *The Question Concerning Technology and Other Essays*, Harper, 1977. Heidegger is no simplistic Puritan environmentalist, but a subtle thinker on the inevitable collision between ourselves and our surroundings.

· AFTERWORD

The opening quotation is reproduced with permission of Henry Holt and Company, New York, from "XLV" in *The Collected Poems of A.E. Housman*, 1945.

The End of Science . . . Addison Wesley, 1996.

a famous saying of Newton's . . . "If I have seen a little further it is by standing on the shoulders of Giants." We are not giants; we are powerful, but not wise.

INDEX

∵

Page numbers that appear in italics refer to Figures.

EOS

Books by Gregory Benford:

"Gregory Benford is a distinguished physicist, astronomer and professor, but first and foremost he is a superb storyteller."—*The Houston Post*

Available wherever books are sold, or call 1-800-331-3761 to order.